D0710617

Lynda V. Mapes

ORCA

SHARED WATERS, SHARED HOME

Lynda V. Mapes

Photography by Steve Ringman, Center for Whale Research, and Others

Graphics by Emily M. Eng

A Co-Publication with *The Seattle Times*

BRAIDED RIVER

The Seattle Times

For the late Chief Tsi'li'xw Bill James of the Lummi Nation

Sunrise 1944, Sunset 2020

With gratitude

CONTENTS

Orcas stay with their families for life, never dispersing—a trait very unusual in nature. *(Dave Ellifrit/Center for Whale Research; taken under NMFS Permit 15569)*

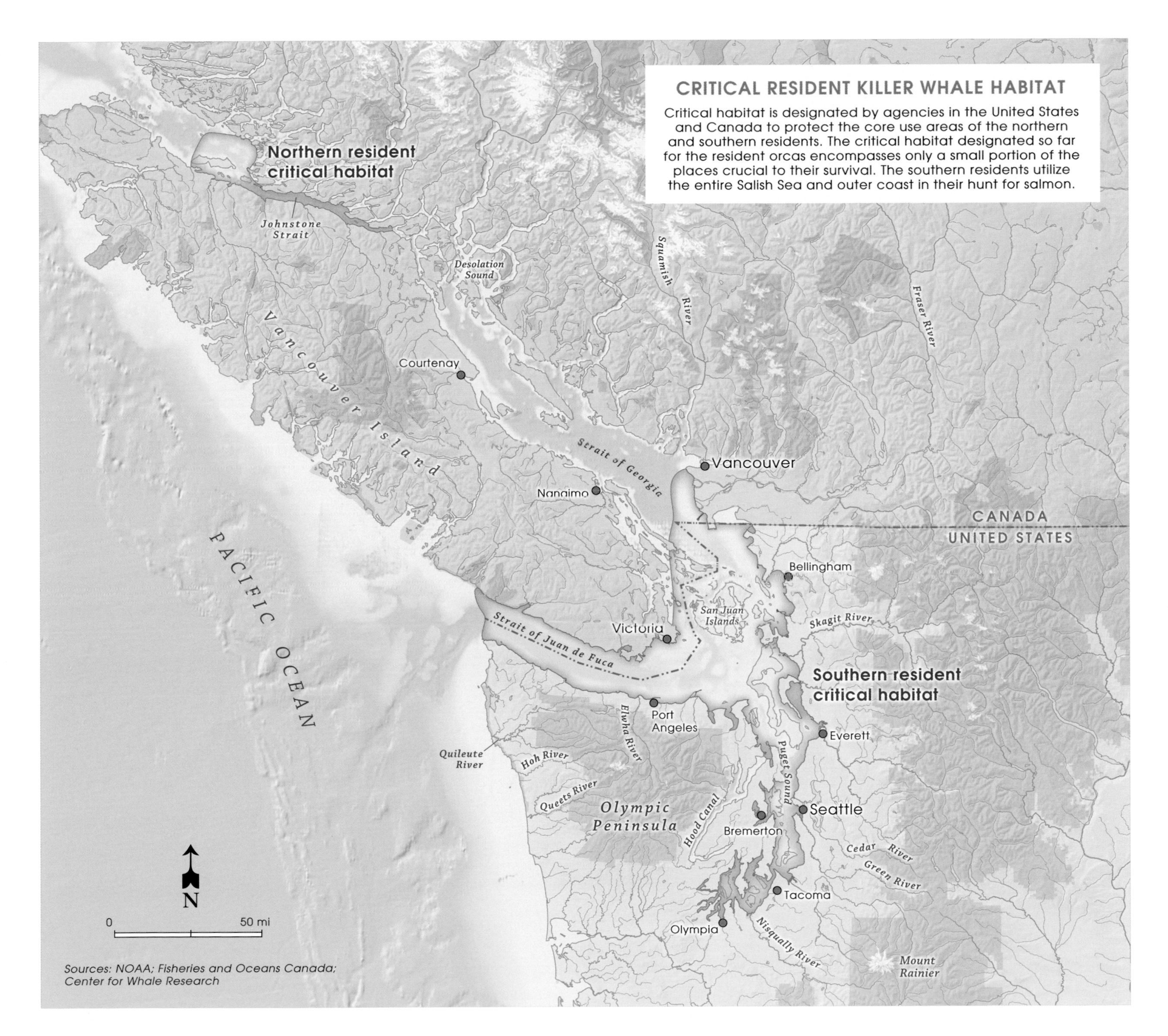

CRITICAL RESIDENT KILLER WHALE HABITAT
Critical habitat is designated by agencies in the United States and Canada to protect the core use areas of the northern and southern residents. The critical habitat designated so far for the resident orcas encompasses only a small portion of the places crucial to their survival. The southern residents utilize the entire Salish Sea and outer coast in their hunt for salmon.
Northern resident critical habitat
Johnstone Strait
Desolation Sound
Squamish River
Fraser River
Vancouver Island
Courtenay
Strait of Georgia
Vancouver
Nanaimo
CANADA
UNITED STATES
PACIFIC OCEAN
Bellingham
San Juan Islands
Skagit River
Victoria
Strait of Juan de Fuca
Southern resident critical habitat
Port Angeles
Everett
Elwha River
Puget Sound
Quileute River
Hoh River
Queets River
Olympic Peninsula
Hood Canal
Seattle
Bremerton
Cedar River
Green River
Tacoma
Olympia
Nisqually River
Mount Rainier
N
0
50 mi
Sources: NOAA; Fisheries and Oceans Canada; Center for Whale Research

PREFACE

Why another book about orcas? Even a casual count on my own shelf numbers about twenty orca titles. So I feel the need to explain this one.

This book started in 2018 with a mother orca as she carried her dead calf, which lived only a half hour. It began as a chronicle of an emergency brought to light by mother orca Tahlequah.

The extinction crisis endangering the orcas that frequent Puget Sound in Washington State is well known, at least in the Pacific Northwest, where these orcas coevolved thousands of years ago with the salmon they rely on. But never before had their struggle been so thrust into the hearts and minds of people around the world.

Tahlequah changed the conversation about these orcas. Her journey of grief, as she carried her dead calf for 17 days and more than 1,000 miles, told us about the state of not only her world but our own. The orcas that frequent Puget Sound and the chinook salmon they primarily prey on—twin monarchs of the Pacific Northwest—both have been pushed to the brink of extinction.

Mount Baker provides a spectacular backdrop for K20, a female born in 1986. Orcas can swim 75 miles a day and more, with bursts of speed up to 30 miles per hour, and are capable of diving deeper than 3,000 feet. *(Steve Ringman/*The Seattle Times*)*

OPPOSITE: Wherever they go, the southern resident orcas draw a crowd of admirers. *(John Gussman)*

RIGHT: The author on assignment for *The Seattle Times* with J, K, and L pods of the southern resident orcas in Puget Sound in November 2018. *(Steve Ringman/*The Seattle Times*; taken under NOAA Permit 21348)*

This book began as a series of newspaper articles and an online presentation of stories, graphics, documentary films, and photographs that explored and explained the roots of the orcas' plight. "Hostile Waters," published by *The Seattle Times* from November 2018 through September 2019, was one of the most ambitious projects ever produced by the *Times* and won national and international journalism awards.This book takes the story further and delves more deeply into it.

For me no story has ever been quite like this one. The articles in *The Seattle Times* about mother orca Tahlequah touched millions of readers. And I think I know why.

Tahlequah's story is about not only the lives of these orcas but our prospects for a shared future with nature and the animals that were here long before us. The story that started with Tahlequah ends up being about us too.

How it turns out is very much in our hands.

INTRODUCTION
TAHLEQUAH

It was the tiniest flash of white: all I could see of the calf born to the orca known as Tahlequah. Day after day, she had clung to her dead calf, refusing to let her baby go.

Banging across the waves in a small open skiff, I was looking for mother orca Tahlequah and her family, scanning the water, ruling out one orca after another: not her. The day was turning into evening, the seas growing choppy and inhospitable. Finally, in a 1,000-foot-deep stretch of water in the Southern Gulf Islands of British Columbia, she came into view, her family traveling around her. She was still holding her dead calf.

The baby could be seen when Tahlequah balanced the calf on her head but was harder to spot when she gripped it in her teeth by one pectoral fin, being ever so careful not to damage the tiny calf she had carried within her

The tiny dead calf born to J35, or Tahlequah, is seen in this July 2018 photo alongside Tahlequah's head as the mother carries her for an eighth straight day. *(Steve Ringman/*The Seattle Times*; taken under NMFS Research Permit 21114)*

Southern resident orcas are forced to contend with the noise, pollution, and industrial development of millions of people. Here Tahlequah pushes her dead calf past the coal docks at the Westshore Terminal at Roberts Bank in Delta, British Columbia. *(Taylor Shedd/ Soundwatch, The Whale Museum; taken under NMFS Research Permit 21114)*

for seventeen months. A female, the baby lived for only about a half hour.

Mourning is well-known in the animal world; giraffes, elephants, dolphins, and apes all have been seen tending, carrying, and staying close by loved ones that have passed. I had heard that orcas do this, too, that mother orcas will sometimes carry a calf they had lost, usually for a few hours. But day after day? This was new even to the experts.

Tahlequah was still within hearing of her family but swimming off by herself. She was near a rock wall where the sunset light glowed pink, a picture-postcard-perfect sight if the scene were not so sad. This was not the happy breaching-orca moment that helped make the waters of the Salish Sea—encompassing Puget Sound, the San Juan Islands, and the waters off of Vancouver, British Columbia—a world-renowned tourist destination. No, this was a reminder of the realities these orcas endure. For while they inhabit these waters in summer, most of the year these orcas—the most urban in the world—share their home waters with the impact of some six million of us, with our coal ports, container terminals, Superfund sites, depleted fish runs, and industrial shipping lanes.

As the sun sank, the wind dropped. The water was still; there were no other boats around. It was so quiet I heard Tahlequah's breathing, labored and primal. She arched her back, diving deep to retrieve the calf, which she had dropped in order to surface and take another breath, before the current could sweep her baby away.

Tahlequah had been doing this around the clock for days, with a 7-foot-long carcass weighing at least 400 pounds, relentless in her unceasing focus. What does it take to do that, day after day, mile after mile, each time having to decide whether to take a deep breath and do it again?

"What is beyond grief? I don't even know what the word for that is, but that is where she is," said Deborah Giles, science director for the nonprofit Wild Orca and researcher with the Center for Conservation Biology at the University of Washington. "She has to prime herself six, seven breaths to take a deep, long dive to go get that carcass," Giles said. "What is killing me is, when is it going to be the last time? And she has to make that decision not to go get it."

Giles and other orca experts were becoming concerned that Tahlequah was endangering herself. A healthy, young female orca at just twenty years old, she is crucial to the future of this family of orcas. By August 2018, there were only seventy-five orcas in this group left; as 2020 drew to a close, there were just seventy-four southern residents left.

In so small a population of endangered orcas, every baby matters. Every orca counts. The orcas that frequent Puget Sound are among the most studied animals in the world, and each orca is known as an individual. Scientists keep a photographic record of the families, with every orca identified by the unique white marking—called a saddle patch—at the base of the dorsal fin, in addition to the fin's shape and any permanent scars and marks. Scientists call each orca by an alphanumeric designation, with a letter for its pod: J, K, or L. There also are nicknames for the orcas. Tahlequah is the nickname for orca J35.

Tahlequah continued her rhythmic diving and surfacing and was close enough that I could see the water sheeting off her sleek black-and-white body—and just a tiny corner of the calf's dorsal fin, so very small. "I am relieved we see her, that she is healthy and swimming strongly and that she is with her family," said Taylor Shedd, as he quietly moved the boat away from Tahlequah. It was hard to leave her as we turned to head back to the dock in the fast-fading light.

Then an educator with the Soundwatch boater education program run by The Whale Museum at Friday Harbor on San Juan Island, Shedd had been keeping vigil with Tahlequah day after day, to keep curious onlookers away—and to help journalists witness what was happening. "It is so emotional that she is so caring," he said. "It boggles my mind. To carry it is hard for her physically and mentally. It is just heartbreaking. She is blowing me away with how fast she can swim and how far she can go."

I wrote about Tahlequah nearly every day in *The Seattle Times* since she began her journey, which continued for 17 days and more than 1,000 miles. Throughout her ordeal, Tahlequah's family remained close around her, taking turns staying by her side. I am convinced she never actually gave up on the calf. It was badly decomposed by the seventeenth day, and I think it finally just fell apart. Afterward, she was seen surging through the water with her family, seemingly no worse for wear. But I knew this was only one of many trials for her family.

Tahlequah had already lost her sister in 2016. That orca probably died of complications from birthing Tahlequah's nephew. He was the next to die, at just ten months. Still suckling his mother until she died, he had no chance without her, despite his family's heroic efforts, including those of Tahlequah, who brought him chunks of salmon. His family members held him up when he was too weak to surface to breathe. When he died, his body was covered with tooth rakes, marks made by the orcas using the only thing they had to hold him up to breathe: their teeth. In the summer of 2019, Tahlequah would lose her mother, too, who gradually wasted away while still in her prime, leaving not only Tahlequah but a son and other youngsters depending on her for survival.

These orcas have been in trouble for years, listed in the United States for protection as an endangered species since 2005 and in Canada under the Species at Risk Act since 2003. Yet it was not until Tahlequah's journey in the summer of 2018 that the orcas that frequent Puget Sound sparked worldwide concern. I think I know why: This wasn't just an animal story. It was a story about a mother who had lost her baby. A family, standing by her in grief, that had already lost so many of their own, caught in a struggle for survival not of their making. Tahlequah's calf was the first live birth to these orcas in three years and a female, at that. Hope had faded in half an hour.

Anyone who had ever lost a family member could not help but be moved by Tahlequah and her plight. A boundary-crashing empathy connected millions of people with a mother who happened to be an orca. She also triggered a deep dread. What was so wrong with the environment that it could no longer support its spectacular signature species? What did that mean for our future, as well?

We asked *Seattle Times* readers to tell us how Tahlequah's story affected them, and more than one thousand people responded. One reader composed a song. Researchers broke down in tears, and elected officials privately said they were losing sleep and unable to function. I received poems and long messages from readers sharing intensely private grief. More than *six million people* around the world had followed Tahlequah's story online in the *Times* by the time she stopped carrying her calf. But the story was far from over.

At the same time Tahlequah was carrying her dead calf, another orca was slowly—and very publicly—starving to death. A spunky little three-year-old orca, known for her spectacular multiple breaches that seemed so full of joy, she was the first in a baby boom of orcas beginning in 2015 that had sparked hope for this population of orcas. Then she, too, was failing.

J50 was nicknamed Scarlet because she still bore the rake marks believed to be from orcas who had midwifed her birth. She was wasting away as the public watched, developing a dreaded syndrome called peanut head. She had become so thin that the normally smooth shape of her body was transformed to a grotesque indent at the neck, with the bones of her cranium showing. Out on the water, Deborah Giles and I smelled Scarlet coming before we saw her. The little orca's breath as she starved had become that foul—no longer the salmon-scented sweet exhalation of a healthy orca but a sour, putrid stench. The smell of death.

Determined to save her, the National Oceanic and Atmospheric Administration (NOAA), charged with orca recovery, mounted an unprecedented international effort, calling in experts from across Canada and the US to implement a detailed rescue plan. The idea was to feed J50 medicated live fish and nurse her back to health in a sea pen if need be, then return her to her family.

TOP: This is one of the last known photos of three-year-old female orca J50, taken September 7, 2018. The so-called peanut head, shown here, is a deformation caused by starvation. *(Katy Laveck Foster; taken under NOAA Fisheries Permit 18786-03)*

BOTTOM: Veterinarian Martin Haulena of the Vancouver Aquarium delivers a dart full of antibiotics to J50. *(Ocean Wise; taken under NOAA Fisheries Permit 18786-03)*

But veterinarians struggled even to find out what was wrong with her. All of these orcas lack regularly available, quality chinook salmon; it is one of the primary threats to their survival. But why was she starving to death? No one had ever seen so skinny an orca still alive. And how to help a wild orca, or even attempt to intervene amid so close-knit a family, without making the situation worse? No one was going to take her away from her mother, that much was decided. But if she moved off by herself somewhere, scientists hoped to intervene.

The Lhaq'temish People, or Lummi Nation—a Coast Salish community in northern Puget Sound,

LEFT, TOP: The southern residents put on a show in their core summer habitat of the San Juan Islands. *(Steve Ringman/*The Seattle Times*)*

LEFT, BOTTOM: Mother orca Tahlequah clung to her perfectly formed, beautiful calf for 17 days and more than 1,000 miles, refusing to let her go. The calf lived only about a half hour. *(Taylor Shedd/ Soundwatch, The Whale Museum; taken under NMFS Research Permit 21114)*

OPPOSITE: A state ferry pauses to let passengers enjoy orcas frolicking nearby. Such sweet slices of life are increasingly rare in the San Juan Islands in summer as the southern resident orca population declines and the whales spend more time elsewhere searching for food. *(Steve Ringman/*The Seattle Times*)*

part of a large group of related peoples of the Pacific Northwest coast living in British Columbia, Canada, and the states of Washington and Oregon—helped with J50's rescue effort. The Lummi regard the orcas as relatives, and they conducted a trial feeding for J50. Tribal members slid adult hatchery salmon overboard to the young orca through a tube mounted at the rear of the tribe's police boat and exiting underwater. The idea was to see if she would take the fish. But the ailing orca took no notice and didn't react at all.

Veterinarians next tried shadowing her in a boat with a petri dish on a long pole over her blowhole to gather droplets from her breath to analyze for disease. Nothing was later determined from that sample. They darted her with antibiotics—not easy from a bobbing boat in pursuit of a free-swimming orca. That didn't help either.

We covered all this in the newspaper, stirring up debate over whether it was right for people to intervene in a wild animal's life. Some said no, others that we had already so altered the ecosystem on which these orcas depend that we had caused her suffering and now had a moral duty to help.

Autumn weather began settling on the Salish Sea, and the little orca and her family were lost to view in the fog for days on end, frustrating any effort to help. Finally, one September day she sank below the waves and never was seen again. The last photo taken of J50 shows her grotesquely malnourished, with her skin, her emaciated body, and the water all around her pocked with hard rain. NOAA launched an urgent search for her body that continued for days. Scientists wanted to do a necropsy, to at least learn what had killed her. But after several fruitless days, the search was called off.

OPPOSITE: Orcas cruise the waters off San Juan Island, a core summer foraging ground for chinook salmon. *(Steve Ringman/* The Seattle Times*)*

RIGHT: Tahlequah's story moved people all over the world. In Seattle, a grieving ceremony was held in August 2018 for mother orca Tahlequah and little orca J50 who perished before age four. *(Lynda Mapes/*The Seattle Times*)*

With Tahlequah's journey ended, her baby gone and little J50, too, people in the region seemed moored in grief as surely as Tahlequah had been. "She was stuck in a loop, and we are all stuck in a loop, too, stuck in doing the same things, expecting to get better results. And it is not working," Giles said. Not for the orcas, not for the salmon on which they depend, not for us. "What we need is going to have to be massive, unheard-of, unprecedented change in order to recover this population."

By the next year, three more orcas had died, and the region seemed to have orca fatigue. *The Seattle Times* asked the public if they still thought about Tahlequah. Just a few hundred people responded, and some said it was time to move on. Death, after all, is part of nature, one reader said. Others said the mother and baby orca were in their hearts forever.

Both are true for me. Yes, death is part of nature, and it comes for us all. But what is happening to the orcas that frequent Puget Sound is no natural disaster.

It is easy to blame these orcas' extinction crisis on the dark time in our history when a generation of their young was taken from their families to do tricks for money in captive live-animal shows around the world. But the capture era ended more than forty years ago—at least in Washington State waters.

What is harder to face is that it is our everyday destruction and pollution of the habitat that supports the orcas—and the salmon they eat—that is the major cause of the orcas' decline. This is distressingly old news. Everywhere salmon have ever thrived, they have been driven to extinction by humans. First in Europe. Then on the east coast of the United States and Canada. Now, in the Lower

48, the Pacific Northwest is the last stand for orcas and salmon alike.

In December 2018 came an unexpected ray of hope: a tiny new baby orca, born to L pod. By May of the following year, yet another calf was born—a female—this time to J pod, the orcas that most frequent Puget Sound. Life carried on. Then in September 2020 came two babies, including a new calf for Tahlequah, healthy this time.

Yes, the Salish Sea is still wondrously alive, and the orcas here and the salmon they depend on are some of the toughest, most resilient species in the world—top predators that have radiated into every suitable habitat since the last ice age. Preserve and restore what they need to survive, and they will take care of the rest.

Tahlequah and her extended family raise this challenge: Can ours be a society that amid its prosperity secures a future for orcas and salmon too? Or will the Pacific Northwest lose its signature wild animals and become a place more like everywhere else?

CHAPTER ONE

The People That Live Under the Sea

The *Tyrannosaurus rex* of the sea, orcas worldwide are devastating predators, ready to rip their prey with a mouthful of conical interlocking teeth. Ruthless, precision carnivores, orcas live up to their species name, *Orcinus orca,* "from the realm of the dead." They get their kill, whatever it is, wherever it is, whatever it takes. Globally, orcas target about 140 species—stingrays,

Top predators in every ocean, orcas take on their world with swagger and power. *(Dave Ellifrit/Center for Whale Research; taken under NMFS Permit 21238 and DFO SARA Permit 388)*

The waters of the Pacific Northwest are home to three types of orcas. Offshores are believed to prey on sharks. Northern and southern resident orcas eat salmon, mostly chinook. Transient, or Bigg's, killer whales devour marine mammals—such as this harbor porpoise about to become a meal. *(Candice Emmons/NOAA Fisheries; taken under NOAA Permit 16163)*

sharks, sea lions, herring, minke whales, baby gray whales, octopi, and much, much more. Their teeth reveal their obsessions. Shark killers' teeth become worn down from taking on rough-skinned prey. Chinook shredders have teeth sharp enough to tear big fish in half with a shake of their head.

With the orcas' cunning intelligence, formidable power, and martial-art moves, no animal on their menu is safe. In Australia, orcas thrash their tails to karate-chop great white sharks and rip their livers out. Orcas chase down, ram, and drown baby gray whales off the coast of California and tear off their lips and tongues. They storm the beaches of Patagonia and snatch baby southern sea lions and southern elephant seals right off the sand. They body-slam and hurl white-sided dolphins through the air in the Salish Sea and herd herring in Norway into terrorized silvery torrents, the better to clobber them with their tails to stun them, then gulp them by the thousands.

In the Antarctic, orcas slash through the water, fast and powerful as packs of wolves. They work together to make waves that wash seals right off the ice, straight into their jaws. Orcas swim bluefin tuna to death in the Strait of Gibraltar, attacking them as they enter and exit the Mediterranean Sea during their breeding migration. Orcas chase the fast-swimming fish for forty minutes at a time, until the tuna are exhausted, then close in for the kill. Alaskan resident orcas have learned to respond to the sound of a fisherman's hydraulic winch sure as a dinner bell, swarming in at the sound of the gear being raised to quickly pick luscious, freshly caught black cod right off the hooks.

The largest species of dolphin (a toothed whale), orcas are—while not abundant in numbers—the most widely distributed mammals on Earth, other than humans. Orcas dominate the food web in every ocean, from the pack ice edges in the Northern and Southern Hemispheres to the equatorial tropics.

Orcas easily swim 75 miles a day, with bursts of speed up to 30 miles an hour. They are big animals: males are typically up to 26 feet long and weigh as much as 12,000 pounds; females are smaller at about 23 feet long and 6,000 pounds. Males can sprout a dorsal fin 6 feet tall at sexual maturity.

They have one of the largest brains of any mammal other than humans—and some of the areas of their brains are larger and more complex than our own, particularly the parts that support empathy and social interaction. This explains a lot. Whenever people say to me, "They are just like us," the first thing that flashes through my mind is, *Don't flatter yourself*.

It is their intelligence and culture that truly set orcas apart from all other mammals, humans included. Whether measured in family loyalty, intergenerational support, or diplomacy, orcas are the superior species. For one thing, they are far and away our elders: In their modern form, orcas have been on this earth for some six million years. While there is only one species of orca, they have evolved into multiple populations across the planet. These are societies of great antiquity, each with its own culture, learned and passed on through generations.

One Species, Different Types

Three orca ecotypes roam the northeastern Pacific: residents, transients (also called Bigg's orcas), and offshore orcas, each with different diets, family structure, and behaviors. Resident orcas are further identified as either northern or southern residents, genetically distinct populations delineated by their primary home range. The different orca types and populations don't interact or interbreed. Conflict—warfare or physical violence—between orcas is rare.

The northern residents are primarily found around British Columbia's Vancouver Island, ranging as far north as Southeast Alaska and south beyond the west coast of Vancouver Island as far as Grays Harbor, Washington. Southern residents primarily inhabit the Salish Sea, in addition to West Coast waters all the way south to Monterey Bay, California.

The southern residents travel in pods: extended families comprised of mothers and their offspring, called matrilines, with multiple generations traveling together. The matrilines are grouped into three pods, J, K, and L, distinguished by calls unique to each pod. Together, the three pods form one clan. The northern residents have similar family structure but are far more numerous and organized into three clans, A, G, and R, with numerous pods within each clan.

Bigg's orcas, or transients, are found in marine waters from Alaska to California, including the Salish Sea and even its interior waters of Puget Sound,

Sea lion and other marine mammal populations have boomed since passage in 1972 of the Marine Mammal Protection Act, providing a steady diet for transient orcas. *(Dave Ellifrit/Center for Whale Research; taken under NMFS Permit 21238 and DFO SARA Permit 388)*

such as Hood Canal. Transients form smaller groups than the residents do, up to ten orcas or fewer, and are more fluid in their membership, including unrelated females and their offspring; they sometimes form larger foraging packs.

Offshore orcas are the most mysterious. They are rarely seen and scantly studied because it is so difficult to observe them in the open sea of the outer continental shelf, where they primarily range. They are believed to feed mostly on fish and even sharks—the latter understanding gleaned by observation that their teeth become quite worn down with age. Sharkskin is so rough it was used by Coast Salish carvers for sandpaper.

Genetic research shows the resident and transient ecotypes in the North Pacific diverged from each other at least two hundred thousand years ago and maybe much earlier. Transients and residents look different, but those differences are subtle, with the dorsal fin of a transient orca more pointed than a resident's. The saddle patch—the unique white marking at the base of the dorsal fin—typically is more open in a resident orca, showing some black amid the white, whereas the transient saddle patch tends to be solid white. Residents are also much more vocal than transients, which are silent hunters stealthily stalking their prey, such as seals hauled out on rocks.

Each of the southern resident pods has its own typical range that changes with the seasons as the orcas follow the migrations of the salmon runs. During the spring, summer, and fall, J pod orcas are the southern residents most often seen in the Puget Sound region, including the inland waters of the Salish Sea in the San Juan Islands. The southern residents, particularly J pod, grace the urban waters of the Seattle and Tacoma areas particularly from mid- to late September through November, when their diet changes from nearly entirely chinook salmon to other salmon species, including chum, the second-biggest salmon in Puget Sound waters. Both K and L pods are more coastal, particularly in winter, though they, too, will be seen in inland marine waters.

Orcas in each southern resident pod spend their time within their own family groups, which are led by multiple matrilines (female orcas and their descendants). Resident orca families in each of the southern and northern populations stay together for life—the youth of both genders never disperse, making resident orca societies probably the most enduring among all mammals. Ask any human parent of a teenager—or, for that matter, any couple—just how well their coexistence is going under the family roof, and the ability of resident orca families

KILLER WHALES, THE OCEANS' TOP PREDATOR

Killer whales (*Orcinus orca*) are the most widely distributed whales in the world and found in every ocean. Three orca ecotypes reside in the waters of the northern Pacific coast. While their ranges overlap, they are not known to interact, and each ecotype is genetically distinct. They have unique foraging behaviors, dialects, diets, and appearances.

THE DORSAL FIN The orcas' most recognizable fin varies in size and shape by ecotype and gender. Males tend to have straighter, more upright fins, with male residents occasionally having a forward slant with a wavy back edge.

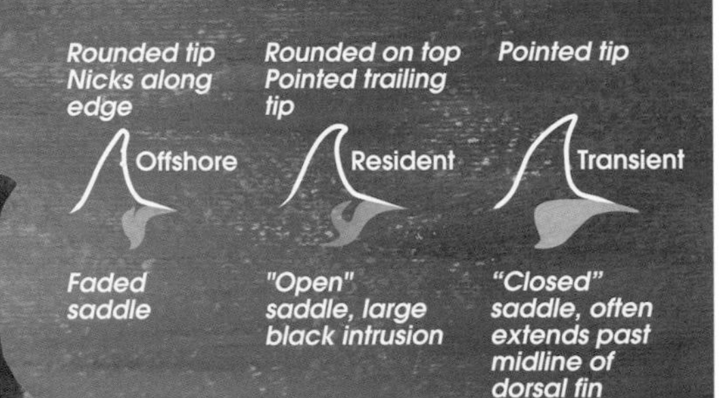

MOBILITY Playful and athletic, orcas can leap from the water in a spectacular behavior called breaching. They "spy-hop" with their heads above water to take a look around, belly flop, roll, and slap their dorsal and pectoral fins. They can travel 75 miles a day and more, with bursts of speed up to 30 mph and are capable of diving deeper than 3,000 feet, about five Space Needles stacked end to end.

ALASKA
Juneau
Ketchikan
BRITISH COLUMBIA
Vancouver
CANADA
US
Seattle
WASHINGTON
Portland
OREGON
Cape Mendocino
San Francisco
CALIFORNIA
Point Conception
Los Angeles
San Diego
PACIFIC OCEAN
MEXICO
0 200 MILES

OBSERVED ECOTYPE RANGES

OFFSHORE
Size: 22 feet
Prey: Sharks
Estimated population: About 300 (as of 2014)
Behavior: Travel in groups of up to 100 individuals

RESIDENTS
Size: 24 feet
Prey: Fish (prefer chinook salmon)
Behavior: Highly social, lifelong matrilineal extended family groups with vocal traditions/dialects

Northern residents
Estimated population: 310 (as of 2019)

Southern residents
Estimated population: 74 (as of 2020)

TRANSIENTS (BIGG'S)
Size: 26 feet
Prey: Marine mammals (seals and porpoises)
Estimated population: About 300 (as of 2013)
Behavior: Smaller family groups—offspring often separate from mothers

Sources: NOAA; Fisheries and Oceans Canada; Center for Whale Research; Esri; Natural Earth

Southern resident adult female orca

SADDLE PATCH Its shape and pigmentation pattern helps identify individuals. The shape may be inherited and helps differentiate between ecotypes.

PATTERNING The distinctive black-and-white coloration varies by ecotype and helps obscure their outline as they hunt, providing camouflage.

6 ft.

At birth, calves are 7 to 8 feet long and weigh 300 to 400 pounds.

LIFE HISTORY In this matriarchal society, female killer whales can live upward of eighty years, while males live fifty to sixty years. A female typically breeds from the ages of fifteen to forty and has four to six surviving offspring, giving birth every three to ten years. The gestation period is up to eighteen months. Babies nurse for about a year; almost half die before their first birthday.

EMILY M. ENG / *THE SEATTLE TIMES*

to stay together through thick and thin, without exception, forever, is bound to impress.

Resident orcas don't switch pods, with rare exceptions, but they will mingle together, including in spectacular gatherings called super pods, when families frolic together. There is even a greeting ceremony practiced by the orcas when they come together after a long absence. The orcas gather in

CLOCKWISE FROM TOP LEFT: J56, still in her first year of life in this photo, swims alongside her mother. Calves of both genders stay with their mothers for life; J50 and her mother, J16, stayed together until J50 could swim no more and sank beneath the waves. Plans to intervene to save the young whale's life stopped short of taking her away from her mother for treatment because orca family bonds are so strong; J27, one of the largest adult male southern resident killer whales, surfacing in the nearshore waters of the San Juan Islands; L121, a juvenile southern resident killer whale, chasing a chinook salmon in the nearshore waters of the San Juan Islands. Resident orcas learn at a young age from their family that food for them is salmon—not marine mammals. *(Dr. Holly Fearnbach/SR3, SeaLife Response Rehab and Research (all photos); Dr. John Durban/SEA/Southall Environmental Associates (top left and bottom right) and NOAA (top right and bottom left); and Dr. Lance Barrett-Lennard/Ocean Wise Research, Ocean Wise Conservation Association (bottom left). Taken under NMFS Research Permits 16163 (top right) and 19091 (all other photos)).*

two lines, facing one another, then erupt into what looks to be an orca party. It's a time for sex and play and catching up, in their way, on the news. The underwater sound of a super pod is jubilant, continuous, and it has the flutelike tones of a calliope, all over the scale.

Summer Splendor in the Salish Sea

The southern resident orcas' summer home has been the Salish Sea for some ten thousand years. It is a place of enchantment, a world of glissading motion, swirling blue and green water, and soft sounds. Out exploring on a midsummer day in 2018, Joe Gaydos, science director of the nonprofit SeaDoc Society, slowed the society's research vessel, the *Molly B*, as we passed by the rocky shore of an uninhabited island. It was just a dot in the San Juan Islands, one of more than 172 islands and reefs, many without even a name, located in the Salish Sea west of Anacortes in Skagit County, Washington. I heard a chorus of quiet peeping.

"Pigeon guillemots," Gaydos said, as the young chicks made soft, high sounds. They were calling to their parents as the adults returned to their nests. Nearby, oystercatchers patrolled the rocks, waggling their brilliant long red beaks, a jaunty contrast with their jet-black feathers. The water was glassy, the sun a golden beneficence. Mount Baker towered supreme over the land and water, more than ten times as high at its glaciered peak as the depth of Puget Sound's darkest abyss. Tidal currents churned and whirled as Gaydos piloted the boat slowly and quietly past more uninhabited islands, tufted with firs and aglow with the brick-red trunks of madronas.

Harbor seals lounged, perhaps sleeping off a meal. Masters of the deep dive, they can plunge to 1,600 feet and stay submerged more than half an hour. They fish by feel, with whiskers ten times more sensitive than a rat's, enabling them to detect undulations caused by the movement of a fish in the water. "They can sense the hydrodynamic track of the fish thirty seconds after it passes," Gaydos said. With their mottled coats, the sleeping seals looked as immobile and inert as the rocks they slept on, hiding in plain sight as they basked. A salmon leapt from the blue, flashing silver-bright, then disappeared into the depths.

Glacier-carved, the Salish Sea is more than 2,000 feet deep in its deepest fjords and straits, but it also shallows to bays and sills, nearshore beaches and sloughs, mudflats and salt marshes, eelgrass meadows and kelp beds. These varied homes nurture thousands of marine species, from whales and seals and salmon to lingcod and rockfish, crabs and clams. In all, about 37 species of mammals, 172 species of birds, 253 species of fish, and more than 3,000 species of invertebrates make their home in the Salish Sea, according to the SeaDoc Society, including some of the world's oldest and largest animals of their kind: Pacific giant octopus and the world's largest barnacle, chiton, sculpin, anemone, and jellyfish. Some Puget Sound rockfish live more than a century, and red urchins abide as long as 150 years. The world's oldest orca was believed to have been southern resident orca J2. She may have been close to one hundred years old when she died in 2016.

Gaydos and I watched as seal pups, just born in the summer height of the pupping season, nursed at the soft curve of their mothers' bellies, eyes half-closed in sleepy delight. The quick arch of a harbor

LEFT: Harbor seals, alert to any movement, are a favorite prey of transient killer whales. This group is lounging on Sentinel Rock off Spieden Island, north of San Juan Island. *(Steve Ringman/*The Seattle Times*)*

OPPOSITE: This orca petroglyph, at Ozette Wedding Rock, documents the long-entwined history of orcas and the region's indigenous people. *(© Brandon Cole)*

porpoise breaking the water's surface punctuated the haze clinging to the blue shine of the water. The porpoise rose and dove, the chuffing of its breath carrying a long way in the quiet. A black, orange, and white rhinoceros auklet hurled its chunky seabird body through the air as if thrown rather than flying. Harlequin ducks ran noisily across the sea, feet slapping the water as they raced for liftoff, the current swirling in whirlpools.

Seals wise to our boat slipped into the water without a splash, their young inching urgently after them. Clumsy on the rocks, they were all liquid grace once they hit the water. The rocks were wreathed with clean, fresh green algae and bronze kelp, and the air had a tangy, briny scent as the tide raised its voice, rushing up the shoreline.

Gaydos turned off the engine and we drifted, the sunshine warm on our backs even as a cool sea breeze came off the water. We started unpacking lunch and talked about all we had seen that morning—including the stark black pinnacles of orca dorsal fins, off in the distance. These highly intelligent, socially sophisticated orca families have undeniable charisma and presence. Some scientists don't like to talk about culture in wildlife, because it sounds like anthropomorphism, Gaydos said, referring to inappropriately attributing human characteristics to the behavior of nonhuman animals. But the southern resident orcas have what humans call culture: learned behavior, passed on intergenerationally, including language, ceremonies, hunting and gathering skills, usual and accustomed gathering and hunting spots, and distinct food preferences.

Cultural Bonds

So much of what resident orcas do is passed on by social learning, said John K. B. Ford, a renowned Canadian orca expert. Beginning in 1977, he was among the first to decipher the calls of resident orcas and understand their importance. Ford discovered there was nothing random about these calls, but rather, they were organized in dialects, taught and learned—a rare thing in the animal kingdom. The ability to learn and repeat sounds is unique to a very small group of animals, just one more way in which resident orcas are so unusual, Ford told me, sharing the thrill of some of those first years of discovery.

"It was a really exciting time; things started falling into place as we began to understand the structure of their society," Ford said. "They are one of the most advanced societies and cultures in the nonhuman world. They are a very special animal; with their dialect structure and acoustic repertoires, they are culturally unique. These dialects have persisted probably for centuries."

The most characteristic signals of orcas around the world are pulsed calls, of about one to two seconds. These are loud calls, audible in quiet conditions farther than 10 miles away. The three ecotypes of orcas—residents, transients, and offshore—produce sounds that are distinct. No orca from any of these different types would mistake who they were hearing.

In addition to calls they share, the resident orca pods each also have calls unique to them. These calls allow family members to travel together in a vast,

dark ocean. They are seldom separated by more than a few miles or hours.

Like the orcas, the Lummi people never stayed fixed in a single place. Regular seasonal migration routes to fishing sites, berry and other harvesting grounds, hunting sites for game and waterfowl, and winter villages encompassed a large territory, sometimes overlapping with other Native American tribes and First Nation bands. The ancestors of the present-day Lummi people occupied much of the San Juan Islands, as well as coastal lands from as far north as British Columbia's Fraser River to as far south as the environs of Seattle.

The Lummi Blackhawk Singers and Tsi'li'xw Bill James, hereditary chief of the Lummi Nation (at left with red headband) lead a procession to a ceremony transferring a piece of culturally important land back to the Lummi Nation. Tsi'li'xw Bill James died June 1, 2020, at age seventy-five. *(Alan Berner/*The Seattle Times*)*

Coast Salish native people have long respected orcas and never hunted them. They see in their lifeways many of their own family traditions, said the late Tsi'li'xw Bill James, hereditary chief of the Lummi people, whose reservation today is located near Bellingham, in Whatcom County, Washington.

Spring rain fell softly, the woodstove ticking with heat as James sat in his favorite chair by the picture window framing a big fir where eagles often roost. His home looked out to the salt water of Puget Sound, toward Lummi Island. I wasn't at all surprised to find students working by the woodstove when I dropped by James's house on the reservation; he was a generous teacher. The students were softening strips of cedar bark to fashion a thick tassel for a woven-cedar chief's hat, traditional regalia of the Coast Salish people.

His mother, the late Fran James, taught him, "Keep your hands busy," and her knitting needles rested on a side table next to his chair. The carved-yew needles, passed down to her from her mother, were polished from years of use. Her homemade spinning wheel, fashioned from scavenged parts, still stood in the workroom curtained off from the sitting room.

The floor was heaped with cedar bark ready for weaving and fragrant, fresh cedar shavings from carving. Baskets and hats in various stages of completion surrounded James, and I knew if I asked, he would bring out the suitcases to show me recently completed handwoven wool blankets and shawls. James was a master weaver, and his work as well as his mother's is cherished by collectors and hanging in museums all over the region.

The quiet, rhythmic thud as two students pounded cedar bark punctuated James's sentences as he began sharing the story of how the orcas, also called blackfish, came to be known by his people as family.

"An elder told me the story of the blackfish and how important it is that we remember who we are and where we come from and what they mean to us," James said. "The Changer that made the human beings also changed humans to animals, and in this way all living things in the Lummi understanding are related. That is how we have to show our respect to all the things that were created from the human beings. We remember, we give thanks to all the things we eat, that we gather. We give thanks because we know they came from our people at one time.

"This is just a glimpse of how we believe," James said. "There are many stories of the people that live under the water, under the ice, the star people—they are all related to us. I know lots of stories, lots of the history that I carry within me. The blackfish are called *qwel lhol mech ten*: 'the people that live under the sea.' There is a long story of how we believe they are related to us." James said:

There was a young man, twelve or thirteen years old, an orphan being raised by his grandmother on San Juan Island. He had nobody to play with or talk with, just his granny who was raising him.

Every day he would wander around with nothing to do. He would sit on the edge of the bluff, looking out over the water. One day he saw *qwel lhol mech ten* was coming.

He was fascinated with the blackfish and went home and told his grandmother. The next day, he saw a young female blackfish again in the group. This time, he saw her jump, and that was good.

Called the side-by-side tribe by the Namgis First Nation, northern resident orcas have the same close family bonds and culture as southern residents. Families, led by the matriarchs of the clan, stay together for life. *(Steve Ringman/*The Seattle Times*)*

Day after day went by, and he decided he was going to jump in his little single canoe and be right in the water when the orcas came. There they came, so close, right like that, and he followed them around for a little while. Then the female jumped high, and they all went down. And down they stayed. He waited and waited. He knew they need air; how come they never came back up? He went home and told his granny.

He went to watch them again, and the same thing happened. Finally, he decided, I am going to be right on their tail and see where they go. The last one went down and sucked him under the water, and he went right behind her in his canoe. He could breathe because of her bubbles and the vacuum in his canoe. He looked and could see the bright light at the top of the water and knew, "That must be where I have to go to get out. That must be where I have to go." He dove deep and followed the blackfish first down—and then out of the water.

He could see them all were coming out on the beach there. They took their skin off, and they were people, people that lived under the water.

The young girl, the last one that was getting out of the water, said, "What are you doing here? You don't belong with us, go back in your canoe. I am going to tell my grandfather."

He came out real strong. "Get in your canoe, go back where you came from," he said. But the boy was afraid the strong current would carry him far from his home. He wanted to wait. And Grandfather said, "Well, I guess you can stay 'til the tide turns and can take you back home."

Grandfather said to the girl, "You watch him so he is not going to get in trouble." She watched him, and she watched him, and pretty soon, the grandfather was watching *them*. They were both teenagers. Pretty soon, little sparks took place.

[James snapped his fingers on both hands.]

The grandfather saw these little eyes meeting each other. The young man asked, "Can I come down here and live with you?" He said, "Just think about it for a minute. . . . Your granny up there on the rocks, she is crying and crying because she lost you to the water. Your granny is up there crying, think about that." The boy put his head down and felt bad. "You go back and you ask her. If you get her permission, you can come down and stay with her."

Away he went, and sure enough, there was granny, just crying, she had lost her grandson to the water. She was so happy to see him when he came back. He told his granny about what he saw, that he saw the people under the water. She said, "I know that already." She already knew.

And so he asked his granny, he said, "Granny, I found a girlfriend down there, and I want to go down there and live with her." She said, "Sure, you have a life now. I am old, I am not going to be here much longer." But he still felt bad. He told his granny, "You go down to the beach every day. Every day I will put the fish there for you. I'll pile the wood there for you. I will take care of you every day." This is what happened.

Every day he put fish for her and wood for her. Pretty soon she was sitting up on the bluff, and then it was time for the *qwel lhol mech ten* to come and she would look. Pretty soon, she saw there were two little ones. "Oh, that must be my grandson." They were gone for a little while, but pretty soon, the day came, and they came back, the two little ones and a little bitty one. "Oh, that must be my great-grandson."

"This is the story of our people and how we know there are people that live under the water," James said. "And this is just one of the many stories I was told. There are many, many creation stories. They were told by our people about who we are and where we come from. We have to share these so our young ones know."

The human and orca cultures of the Salish Sea have shared these waters for thousands of years. And like the Lummi and other Coast Salish communities, the southern resident orca families share customs, culture, language, a deep knowledge of the water, and food.

Just as in Coast Salish communities, the role of the matriarch is also crucial in orca families. Orcas are among only five species of mammals on Earth, including humans, in which the female has a long postmenopausal life stage. "The question is why, and it really boils down to the family structure that these animals, and early humans, evolved with," said Darren Croft at the University of Exeter in the UK, a coauthor of a 2019 paper published in the Proceedings of the National Academy of Sciences revealing the reliance of even adult orca sons on their mothers. "Food sharing was very important for early humans, and you have generations living together. The older generations can increase survival of the younger generations. Their landscape is a social one; they are taking the opportunity to increase the survival of their relatives."

It's the same for the southern residents. In addition to helping the pods find food, the grandmothers assist with caring for the young and

LEFT: Southern resident orcas are a thrill whenever they return to local waters. This orca and its family were in Puget Sound just north of Seattle in the fall of 2019. *(Steve Ringman/* The Seattle Times*; taken under NOAA Permit 21348)*

OPPOSITE: The southern residents are endangered because of three primary threats: lack of adequate chinook salmon, pollution, and noise and disturbance from vessels and boats that makes it harder for orcas to hunt. *(Ken Balcomb/Center for Whale Research; taken under NMFS Permit 15569)*

instruct in maternal care. The matriarchs, who can live eighty years and longer, have been through many seasons, and when it becomes more difficult to find food, it is the matriarchs that lead the way. "The amount of salmon in a given year varies, and that kind of information is knowledge acquired over time with age and experience. It is the old females, the postreproductive ones, that are out in the lead," Croft said. "Guiding the group, they do this more when salmon is in short supply, when they know the family is struggling and not finding enough food. That leadership behavior gets pretty extreme."

Behavior ecologist Michael Weiss, a coauthor on the 2019 paper, did his field research that revealed the so-called grandmother effect at the Center for Whale Research on the west side of San Juan Island. It was a stakeout every year as researchers waited for the orcas to come back to their summer foraging grounds. Particularly in scarce salmon years, the wait stretched later and later into the year. And inevitably, it was the oldest orca of all the southern residents leading the way: J2, called Granny.

"J2 was a very old orca, and she was traveling sometimes a mile ahead of the group, still in vocal contact. It was so interesting, my first summer on the water—I go out, and the first orca, I only see one, and I know it is her," Weiss said. "When you talk about culture, there really is good evidence for it. She was very old, probably in her nineties. She was maybe the oldest orca in the world." Yet Granny was still nurturing her relations, even to her own detriment. J2, who disappeared in December 2016, was visibly thin in the last known photo of her. In it, she was sharing a fish.

A Dwindling Population

Prior to the twentieth century, there may have been more than two hundred orcas in the southern resident population, according to the federal Marine Mammal Commission. But the persecution of the orcas by fishermen and later the live-capture industry, combined with depletion of chinook runs in the Columbia River and elsewhere, devastated the pods. By 1974 the southern resident population sank to just seventy-one animals.

The J, K, and L pods entered an eleven-year growth period peaking at ninety-eight orcas in 1995. Then from 1996 to 2001 their population crashed by almost 20 percent to just eighty individuals before expanding to eighty-eight orcas in 2005. The steep drop and modest rebound prompted listing of the southern resident orcas for protection under the federal Endangered Species Act in 2005. With long life spans, low reproductive rates, and only a small number of reproducing orcas in the population, the southern residents are even more vulnerable to extinction than numbers alone would indicate.

Genetic analysis has shown that the largest and oldest males are the primary breeding orcas in the southern resident pods, with just two orcas siring most of the offspring. Paternity testing has also shown that southern residents sometimes mate with orcas in their own pod and even in their own family, potentially leading to inbreeding problems, on top of their other troubles.

Scientists are continuing to investigate the threats to orca survival, from inbreeding and disease to lack of prey. Pollution also poisons their bodies—especially baby orcas, which get the biggest dose in their mother's milk. Vessel noise and disturbance even by small nonmotorized boats,

Southern resident orcas need abundant chinook salmon, in waters quiet enough to successfully hunt them, throughout their foraging range. Like people, orcas are not adapted to fasting and need to have enough to eat every day in order to be in prime health at every life stage. *(Candice Emmons/NOAA Fisheries; taken under NOAA Permit 16163)*

such as kayaks, interrupt orca foraging and mask the sounds they need to hear in order to hunt. These threats are interrelated. When the orcas go hungry, they burn their fat, where toxic pollutants accumulate, thus adding to the toxics circulating in their blood. Noise is a problem because it compounds the problem of the scarcity of chinook: noise and disturbance make scarce fish even harder to find.

Preferred Prey

Orcas in both the northern and southern resident populations prey primarily on chinook salmon—the largest of all salmon, delivering the most calories for the hunting effort. In the northeastern Pacific, for the thousands of years during which the resident orca populations evolved, it made sense to specialize in catching chinook, because these large fish were abundant and available year-round.

In some of his research of prey samples of resident orcas in summer, orca expert John K. B. Ford found that they were eating more chinook than anything else, even when other salmon such as sockeye and pink were far more abundant, outnumbering chinook by as much as five hundred to one. Chum are also a significant source of food for resident orcas, particularly in the fall. But chinook are still preferred, when they are available. Subsequent research has confirmed that, in winter, as much as half the southern residents' diet is coho and chum, even a little steelhead, and some lingcod, skate, or flatfish. What these predators need the most, however, is chinook. To stay healthy, an adult orca must catch about eighteen to twenty-five salmon every day, or up to 300 pounds, depending on the age and condition of the orca. Lactating females need even more.

Chinook are deep-diving fish, good at hiding in holes, and the deeper chinook go, the darker it gets. But that's no problem for an orca on the hunt. Water is a terrific medium for sound, which can travel more than four times faster in water than in air. In dark or murky water, echolocation is a much better tool for hunting than sight.

Seeing with Sound

Orcas have no vocal cords—the mouth is not the source of the sounds orcas make. Instead, an orca uses phonic lips on either side of its blowhole as deftly as a horn player, shaping air flow with those lips to make sounds.

Orcas also can search for fish, or good fish habitat, using echolocation clicks: bursts of sound focused through fat in a reservoir called the melon, located at the front of the head. The echolocation beam is not fixed; an orca can flex its melon to point and focus a click train of sound. As an orca homes in on a fish, it makes faster clicks and urgent calls until an echo from the clicks bounces back: it rebounds from the body of a fish.

Another fat-filled reservoir in the orca's lower jaw receives the echo and conducts it to the middle ear, inner ear, and then hearing centers in the brain via the orca's auditory nerve. In this way, an orca uses echolocation to actually see *inside* its prey, to the swim bladder of a fish. This gas-filled sac in the fish maintains its buoyancy. Orcas know by the size and shape of the swim bladder just what species of fish they have targeted—specifically, whether it is a chinook. With a bead on its favorite prey, the orca's clicks change to a rapid buzz. It's a surprising sound, as if from an insect rather than a 6-ton mammal.

It can be hard for humans, so sight-oriented, to grasp how essential sound is to the southern resident orcas, said Marla Holt, a research wildlife biologist with NOAA's Northwest Fisheries Science Center in Seattle. But sound is critical to all of the orcas' essential life functions, from sticking together in a dark ocean to finding mates and hunting fish.

In much of her work, Holt seeks to help people understand the *umwelt* of the orca—its sensory world. Researchers use a long pole to tap into place a sensor that sticks to the surface of an orca's skin, allowing them to eavesdrop on hunts. These multipurpose sensors capture not only the clicks, buzzes, and calls of an orca on the hunt but also the orca's movements and the sounds in its environment. Scientists are learning from data collected with these sensors how orcas maneuver in the darkest waters to chase down fleeing chinook salmon, one at a time.

Salmon probably know when they have been targeted; pressure of the sound waves from an orca's echolocation clicks may be felt by the fish tactically, along its lateral line, a nerve that sensitizes its sides. "You can hear the fish respond by diving deeper; they escape to the bottom—and the whale follows them," Holt said. "It's a prolonged chase. It's a lot of effort for the whales."

For a southern resident orca, the average maximum prey-capture dive is about 400 feet deep

ECHOLOCATION

Toothed whales, including orcas, and most bats have the ability to locate and identify objects through echoes, which are reflected sound. For killer whales, echolocation is crucial for hunting salmon.

HOW IT WORKS

1. Breathed air enters the blowhole, goes through the nasal passage, and fills the first set of air sacs. During underwater dives, a nasal plug closes the nasal passage to the blowhole.
2. Below the top air sacs in the narrow nasal passage are phonic lips on each side of the blowhole. The whale uses surrounding muscular structures to manipulate air flow that causes the phonic lips to vibrate, resulting in acoustic pulses that sound like clicks.
3. The clicks pass through the melon, an organ at the front of the whale's head, made of specialized fats. The whale can change the shape of the melon and focus the sounds into an acoustic beam it uses to scan its environment, like a flashlight.

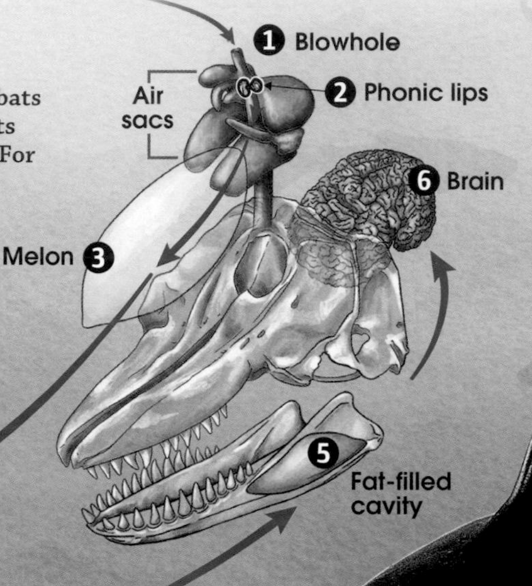

4. The sound wave bounces off objects in the whale's path and returns as an echo.
5. The echo enters through the thinner end of the lower jaw where a fat-filled cavity is located. The echo is conducted to the middle ear, inner ear, and auditory nerve before finally reaching the hearing centers of the brain.
6. Before the next click is produced and sent out, the echo of the previous one must be received. These echoes not only describe the distance of an object but also its size, shape, structure, composition, speed, and direction. Killer whales can distinguish different species of salmon by "seeing" inside the fish, detecting the size and shape of the salmon's swim bladder.

ECHOLOCATION MASKING

Human-caused noise limits the horizontal detection ranges of killer whales trying to echolocate chinook salmon. The shorter the detection range, the less likely the whale will locate and capture the salmon.

THE CLOSER A SHIP IS, THE LOUDER THE NOISE IS.

Ambient Haro Strait

Echolocation-detection range*: 1,300 feet — *0% reduced range*

Container ship moving at 21 knots

Distance: 656 feet away
Detection range: 66 feet — *95% reduced range*

Distance: 1,450 feet away
Detection range: 197 feet — *85% reduced range*

*Echolocation click frequency of 50 kHz at the surface seeking a chinook 213 feet below.

THE FASTER THE BOAT GOES, THE LOUDER THE NOISE.

Ambient Haro Strait

Echolocation-detection range*: 1,300 feet — *0% reduced range*

Boat 328 feet away from the whale**

Speed: Cruising at 24 knots
Detection range: 66 feet — *95% reduced range*

Speed: Below 24 knots
Detection range: 164 feet — *88% reduced range*

**A 29-foot aluminum monohull boat with twin 225-horsepower outboard motors.

Source: Marla Holt, NOAA Fisheries Northwest Fisheries Science Center

EMILY M. ENG / *THE SEATTLE TIMES*

and four minutes long. But the orcas will also dive to nearly 1,000 feet, sprint in bursts as fast as 30 miles per hour, and stay down as long as ten minutes, Holt and Jennifer Tennessen, biologists at the Northwest Fisheries Science Center, found in their research published in 2019 in the *Journal of Experimental Biology*. The southern residents don't always dive to the depths for a meal; some hunts start and end right on the surface. But wherever they capture their food, adult females and younger orcas break up most salmon and share it at the surface.

The resident orcas' penchant for a particular food—salmon—is a learned behavior that is typical of orca society. In every ocean of the world, orcas target specific prey they have learned to hunt, using local environmental features and seasonal patterns in the waters they dominate.

During the chinook crash in the 1990s, orca expert Ford and other researchers were puzzled that the northern and southern residents didn't just switch prey when their main foods were scarce. "We started to realize these animals were selecting for chinook far more than we would expect them to," said Ford, "especially in the summer months, when sockeye and pink just were not a significant part of their diet. Why do they persist in looking for a species of prey that is inadequate in abundance to their needs?"

He chalked it up to a kind of cultural inertia. "They evolved as chinook specialists when there was a great abundance of chinook—a far greater abundance—in the Columbia, the Sacramento, the Fraser. It was easily a very realizable food source year-round, not like sockeye that just come in the summer months. It was a prey source they could specialize at capturing, a very reliable and rich food source. It also is a very ingrained, deep-set trait. Young orcas grow up learning what food is; they are provisioned by their mothers. It's not genetically determined; they are a blank slate when they are born, but they learn what food is."

On the Hunt

Prey specialization also is how orcas get so good at what they do, and it shapes a lot of how orcas live. Their prey determines their hunting style.

While fish-eating orcas use sound to hunt, transient orcas hunt marine mammals using stealth, sneaking up silently on their quarry. Marine mammal–eating orcas appear to look for their prey and listen for their sounds rather than use echolocation—it would be a sure giveaway to their pinniped and cetacean prey, which have excellent hearing. Background boat noise can even be a plus for them, covering the sound of their approach.

The need for stealth also drives transient orcas to hunt in small groups. They will hunt cooperatively to take down big prey, such as a Steller's sea lion weighing more than 2,000 pounds—which has big teeth that can maim or even kill. A group of transients will encircle the sea lion so that it can't get to the shore while individual orcas take turns ramming it. This can go on, Ford reports, for hours until the sea lion is sufficiently beaten up to be safely grasped, drowned, and shredded among the family. Transients will also run down even fast prey such as Dall's porpoises by chasing them to exhaustion and will herd Pacific white-sided dolphins by the score into confined shallow bays to pick them off.

To hunt salmon, resident orca matrilines spread out, often over a wide area, with individuals and small subgroups diving and surfacing independently, generally while swimming in the same direction. This is when their calls are so important. Southern residents are believed to coordinate their movements and maintain contact with loud underwater calls that can, in quiet conditions, be heard for miles. When in coastal inlets, channels, and straits, individuals and small maternal groups usually work the shoreline, whereas mature males more often forage alone, diving deeper waters. Ford reports that orcas can detect chinook from more than 300 feet away.

The seasonal movements of their prey drive the movement of resident orcas as they follow the migratory paths and timing of chinook populations. Resident orcas don't just chase any fish anywhere; they congregate during summer and fall in the feeding areas where geography and tidal currents work to their advantage, concentrating their migratory

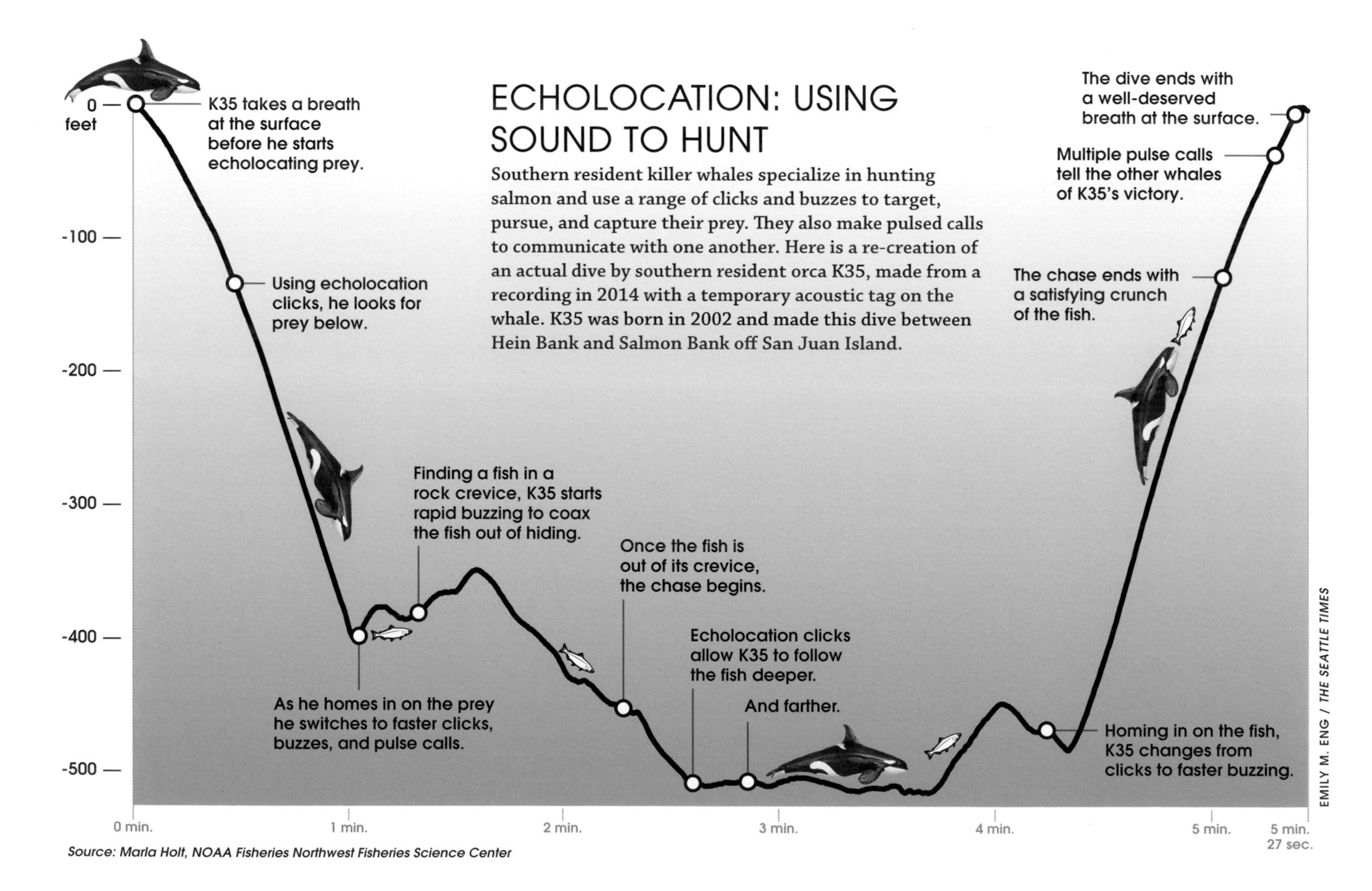

Resident orcas have evolved to hunt and target chinook, the biggest salmon, which reward them with the most calories for the hunting effort. Food preference is a key attribute of orca society and culture in every ocean. Each type of orca specializes in its prey and hunting strategy, utilizing the characteristics of its habitat, and passes this learning and culture on to the next generation. *(Candice Emmons/NOAA Fisheries; taken under NOAA Permit 16163)*

prey. While they travel widely, resident orca pods and matrilines have favorite spots they return to, more often than other types of orcas. These are the places where the mother orcas have learned the fish are, from their mothers before them. It is also why when humans pile on those same places, making it harder for orcas to hunt or depleting the salmon orcas have learned to depend on, orcas suffer.

Unlike resident orcas, which have their favorite spots and family hangouts, transient orcas live up to their name, never remaining long in any one location. They are nearly constantly on the move, swimming from one prey hot spot to another as necessary once prey is alert to their presence. Transient orcas cover more than 90 miles a day. Socializing and resting are seldom witnessed among transients.

Also in contrast to residents, transients in the inner-coast subpopulation stick predominantly to nearshore waters year-round, because that's where their prey is. Harbor seals, harbor porpoises and Dall's porpoises, and Steller's sea lions are nonmigratory and always on the menu. Transients do follow the seal pupping season, with their numbers in nearshore waters increasing in summer, particularly for the easy pickings of pups on beaches or rocks. A pup is a juicy bite for a hungry orca. The outer-coast subpopulation of transients forages on the outer continental shelf and likely feeds more on elephant seals, fur seals, and porpoises.

A Sense of Place

The southern resident orcas are a key indicator of the health of the lands and waters of the Pacific Northwest. "Biologically, species diversity tells you a lot about the health of the ecosystem," the SeaDoc Society's Joe Gaydos said as we sailed on, surrounded with lives grand and small, winged and finned, shelled and furred. "What is lost to everyone's sense of well-being if people start to see this as a dying place?

"The orcas in our backyard, the crabs you caught for dinner—it's that sense of place that you want to take care of. Seattle used to be a frontier town, and while it isn't that anymore, it's still a place where the southern residents swing by. You don't want to lose that. Having them come back every year is magical. It's about feeling this is home."

The southern residents that frequent the Salish Sea are struggling to survive today, in part because they are still recovering from earlier losses. Those happened during the capture era: a time not long ago when orcas were hunted, trapped, and sold in a profitable worldwide aquarium trade. For more than a decade, Puget Sound was the primary source of supply. Southern resident orcas were the main target, particularly the youngest and smallest—the easiest to train and cheapest to ship.

Orcas embody the wonder and beauty of the Pacific Northwest. *(Mark Malleson/Center for Whale Research; taken under NMFS Permit 15569-01 and DFO SARA Permit 388)*

CHAPTER TWO

Captives

Ben Helle needs to use only one search word to find examples of orca dread and mutilation: "blackfish." Consider the front-page story in the *Olympian Daily Recorder* on November 8, 1910, cheering the bravery of two teenage boys for cornering a young orca trapped in shallow water. They shot its eyes out and cut the young orca apart with a knife. It took the orca three hours to die from the boys' knife slashes to its

The capture era in the 1960s and 1970s forever transformed our understanding of orcas, which for centuries had been feared and reviled. *(Ken Balcomb/Center for Whale Research; taken under NMFS Permit 15569)*

throat and uncounted plugs from their .22 rifle. That was just one of many stories Helle, an archivist at the Washington State Archives, dug up to show me how the newcomers to these lands and waters felt about orcas "stealing" "their" fish. They were vicious, *killer whales*, vermin to be avoided or shot on sight.

In 1956 the US Navy used orcas for target practice in Icelandic waters. The Canadians mounted a .50-caliber machine gun overlooking Seymour Narrows, midway down Vancouver Island's east side, in 1961 to mow orcas down. Ultimately never fired, it nonetheless symbolized the war on marine mammals underway for decades around the world. In the Pacific Northwest, the carnage was wholly embraced by state and federal agencies. Until the early 1960s, the State of Washington paid a bounty on every harbor-seal nose mailed to Olympia, and federal researchers routinely harpooned orcas and cut them open to learn what they were eating.

It wasn't until people saw orcas up close—in captivity—that minds were changed forever and orcas went from being reviled to adored and, ultimately, protected. It all started with one man and one orca: Ted Griffin and Namu.

Namu

Griffin was a prodigious collector who wanted the ultimate prize for his waterfront aquarium in Seattle—an orca. In 1965 he got the phone call he had been waiting for: two northern resident orcas had been accidentally caught in a fisherman's net. Word had gotten around about the collector in Seattle who wanted an orca. The fisherman called Griffin and told him to bring cash. Griffin quickly raised $8,000 to buy one of the orcas (the other had escaped). Nearly broke at the time, he got the money mostly from waterfront businesses, including the restaurant and fish bar Ivar's Acres of Clams. They, too, were hoping to cash in if the orca was a hit.

Griffin sat with me on his living room couch recently and quickened with excitement as he told me of his race by seaplane to a cove in British Columbia to get the orca, a backpack full of cash slung over his shoulder. His quest was about more than collecting a moneymaking attraction, said Griffin, in his eighties when he talked with me. He wanted to know the ocean's feared and loathed predator for himself. "The world is confused about the whale," he said, recalling the time when orcas were so detested. "To me, he is just another pet, somebody to make friends with. In my mind, I had already accepted the whale as a companion. And a friend."

Griffin was awkward and struggling in his early school years, and it was with animals that he found refuge and delight. His pets included everything from a seagull to a lungfish to an otter—kept in the house. Griffin built an 8,000-gallon fish tank behind his family's Bellevue, Washington, home. He was scuba diving with rudimentary equipment before he was driving. Flipping through old photos and newspaper clippings, he showed me a picture of himself walking around Green Lake Park in Seattle side by side with a gibbon monkey, Griffin holding the monkey's hand like a small child's.

He started collecting animals for sale early on. One of his first business ventures was a pet shop on Seattle's Aurora Avenue. He opened his aquarium—no relation to the Seattle Aquarium of today—in 1962 on the downtown waterfront, hoping to capitalize on the upcoming World's Fair. Now all he needed was the ultimate marine-park attraction.

TOP LEFT: Namu swims in Rich Cove, near Bremerton, where *Namu, the Killer Whale* was filmed. Ted Griffin and Namu spent day after day together at this cove. *(Bruce McKim/ The Seattle Times)*

BOTTOM LEFT: Ted Griffin makes world history with Namu in 1965, becoming the first person ever known to ride a killer whale in the wild. *(The Seattle Times archives)*

RIGHT: Ted Griffin holds a fish for Namu in May 1966 at his Seattle Marine Aquarium. Namu was the world's first captive performing killer whale. *(Richard Heyza/The Seattle Times)*

Arriving at the remote fishing town of Namu in British Columbia, Griffin set to work making a sea pen for the nearly 400-mile tow to Seattle. Then it was time to get the orca, still confined behind a net, into the pen. This, Griffin decided, was his job.

He put on a wet suit, mask, snorkel, and scuba gear and jumped in the water. "I dive down and, oh God, there is this shadow four feet away, looking at me," Griffin said, remembering his first moments with the orca he named Namu. Then he heard it: a loud *squeak*.

"I think, 'It's the whale,' so I go '*Eeee*' and within half a second, the whale squeaks," Griffin said, his eyes still wide with the memory. "My God, I am crying, I can barely keep my mask on. It is indescribable. What has happened is that all those years I am wanting an animal to say hello, and one has. I am thunderstruck."

The journey was soon underway, with a tug pulling the pen illuminated with kerosene lanterns along the top rail. At first Namu was followed day and night by a female orca and baby, calling constantly and coming within feet of his pen—probably Namu's family—until the tow reached the cultural boundary of the northern residents' territory, midway down Vancouver Island. There, at Seymour Narrows in Johnstone Strait, the two orcas finally turned around and headed back north. Namu was on his own.

As Griffin hauled the penned orca south, word of Namu quickly spread, in no small part because Griffin drummed up publicity with the two local papers, the *Seattle Post-Intelligencer* and *The Seattle Times*. By the time the tow reached Deception Pass in Puget Sound, thousands of people were on the bridge over the pass, hoping to catch a glimpse when Namu's Navy, as the orca's traveling entourage of onlookers, press, and promoters was called, passed beneath. It must have been quite a sight as the fishing boat pulled the homemade sea pen through Deception Pass with, of all things, an orca in tow.

Arriving in Seattle on July 28, 1965, Griffin was given a hero's welcome and a key to the city. A Dixieland band played, and the lieutenant governor of the state and mayor of Seattle declared it Ted Griffin and Namu Day. There were Namu buttons, T-shirts, and even a new dance and pop song named for the orca. The first Sunday that Namu was put on display on the Seattle waterfront, some five thousand people paid to come see him. There were more than one hundred thousand paid admissions in the first five weeks, historian Jason Colby of the University of Victoria writes in his book *Orca: How We Came to Know and Love the Ocean's Greatest Predator*. As Namu's cage was towed to the aquarium, Griffin said he cared about only one thing. "They are making music, celebrating, the fireboats are all shooting in the air. But for me, all I saw was the whale."

No one knew then that an orca could be safely approached at all. But not long after Namu arrived in Seattle, here was Griffin first touching Namu on his face, then his blowhole. He discovered the orca liked his skin scratched with a stiff brush—belly, back, everywhere. They squeaked back and forth, Namu adjusting his tone to match Griffin's. Within a month Griffin made history, becoming the first human ever known to *ride* an orca. Namu, Griffin said, quickly figured out how long Griffin could hold his breath underwater and timed his dives accordingly. The public ate it up.

"Namu the killer whale has been broken to the saddle," wrote *Seattle Times* reporter Stanton Patty

Ted Griffin, in October 2018, shows a *National Geographic* photo of himself prying open Namu's mouth to show off his teeth. The photo was taken by Flip Schulke. *(Steve Ringman/*The Seattle Times*)*

on October 10, 1965, after visiting Namu's sea pen in Rich Cove, temporary quarters for the orca during filming of the feature film *Namu, the Killer Whale*. ("Make Room in Your Heart for a 6-ton Pet" was the film's subtitle.)

"Ted Griffin rode his 8,000-pound pet yesterday afternoon at Rich Cove in a wild, watery rodeo scene," Patty continued. "Hollywood cameras were rolling for a motion picture starring Namu. . . . But Griffin's daredevil performance atop Namu was more than anyone had bargained for. 'He was gentle—extremely gentle,' Griffin said. 'Namu didn't make a malicious move. It was just like riding a ten-cent horse at the supermarket.'"

Griffin and Namu spent day after day together, playing in the cove. "He doesn't want me to go back to shore and go home, as long as I am in the water grooming him or riding him," Griffin remembered. Sometimes Namu would even hold Griffin close in his pectoral fins. "Namu holds me hostage for his pleasure, as I have held him captive for mine," Griffin wrote in his autobiography, *Namu: Quest for the Orca*.

Visitors and the press were crazy for the story of Griffin and Namu. *National Geographic* magazine published a photo of Griffin opening the orca's jaws that Griffin keeps in a frame on the hearth of his home.

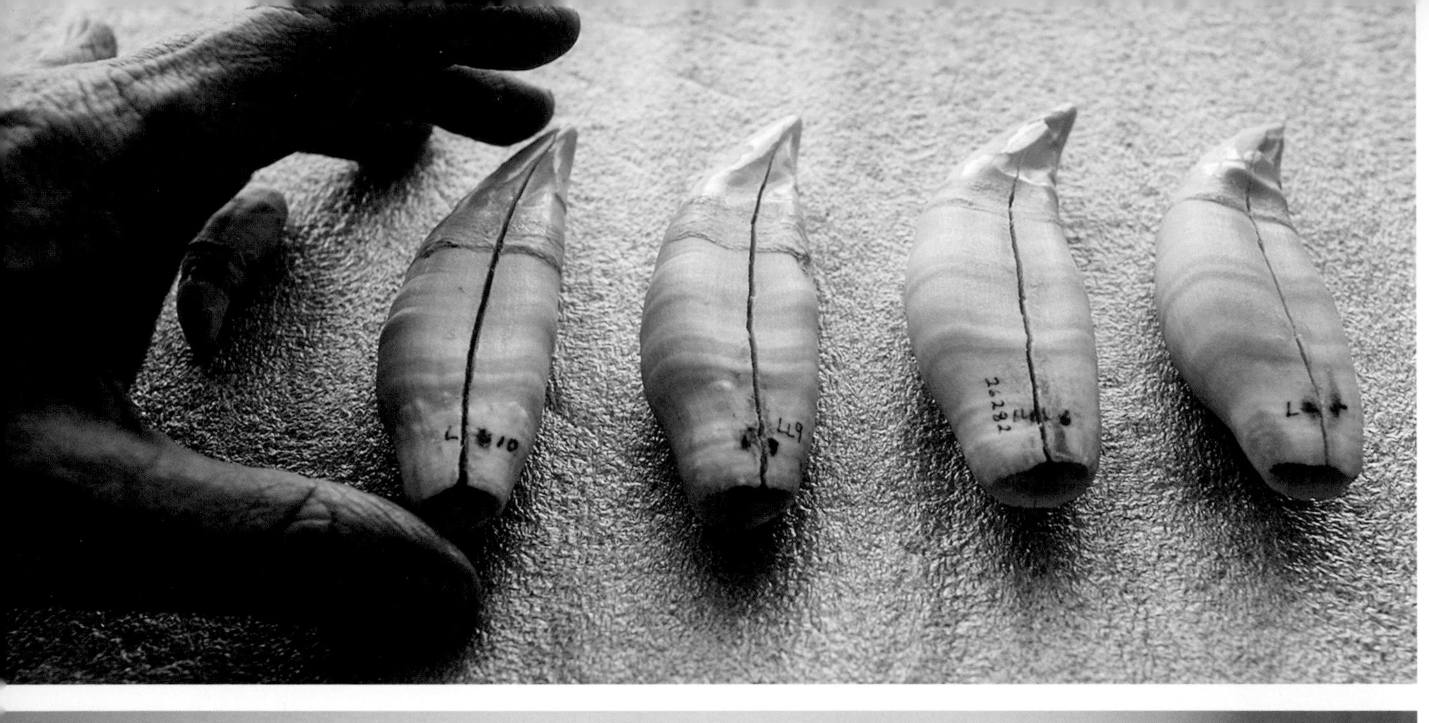

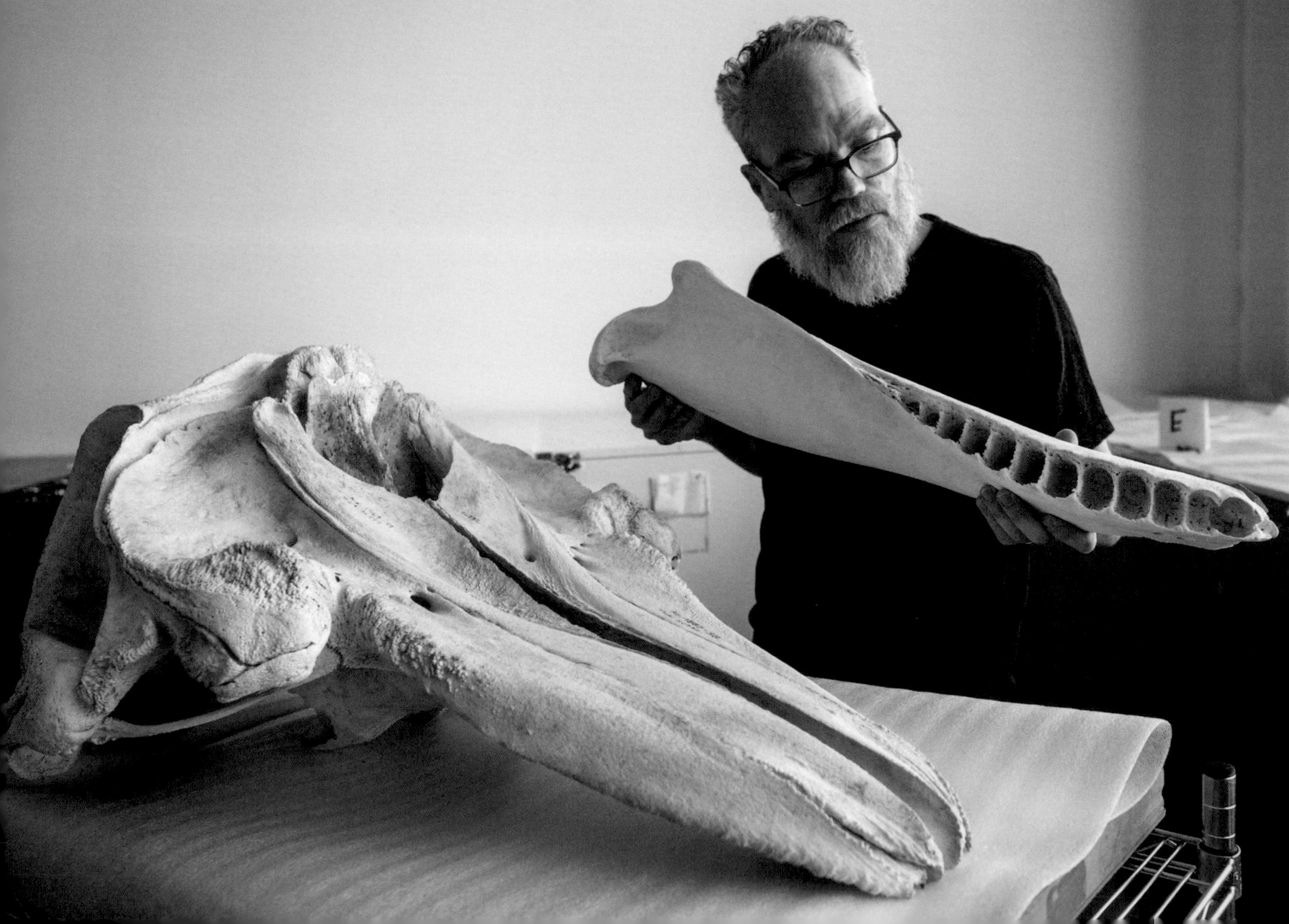

TOP: The teeth of southern resident killer whales are up to 4 inches long, hard and dense as marble, and needle-sharp for shredding salmon. These teeth belonged to Namu and are kept at the Burke Museum of Natural History and Culture in Seattle. *(Steve Ringman/The* Seattle Times*)*

BOTTOM: Jeff Bradley, mammalogy collections manager at the Burke Museum of Natural History and Culture, holds a jawbone from Namu. The big sockets in the jawbone held his teeth. *(Steve Ringman/*The Seattle Times)

Orcas for Sale

Namu fever stoked an international craze for orcas to be put on exhibit. The capture era that would claim the lives of so many orcas—and forever change our understanding of them—was underway. By 1976 some 270 orcas had been captured—many of them multiple times—in the Salish Sea. At least 12 of those orcas died during capture, and more than 50, mostly Puget Sound's southern resident orcas, were kept for captive display. All are dead today but one: Lolita, the last southern resident still alive after more than fifty years in captivity. Only Corky, a female northern resident captured in British Columbia's Pender Harbour on December 11, 1969, at about age four, has been in captivity longer, at SeaWorld in San Diego, California.

I didn't understand until looking through old clippings in *The Seattle Times* the normalcy with which all this was regarded at the time—even when Griffin was hunting orcas. In one photo published in the newspaper on October 31, 1965, Griffin is leaning out of a helicopter and taking aim at an orca just feet below him, with a harpoon gun at his shoulder, ready to fire at the orca fleeing for its life. The photo caption stated blandly, "A pod of killer whales rolled beneath a hovering helicopter while Ted Griffin prepared to fire a tranquilizer dart."

Back then, Griffin was still learning how to hunt orcas, and the *Times* was rooting for him: Griffin hoped to "scoop up a mate for Namu," the newspaper reported, erroneously telling readers "if a tranquilizer harpoon hits a whale it is wounded only slightly." Researcher Merrill Spencer would learn differently when, on a hunt of his own, he inadvertently killed an orca he was attempting to take captive with a tranquilizer gun. Orcas are active breathers; they cannot lose consciousness. They always keep partly awake, in order to keep breathing. The dead orca was hauled to a rendering plant.

As Griffin continued learning how to capture wild orcas, he also intensified his relationship with Namu. He acknowledged his obsession in his autobiography, stating that he was spending more time with Namu than with his own family.

It was all over in less than a year. The summer after he was captured, Namu died a terrible death from a massive bacterial infection caused by the raw sewage polluting Elliott Bay.

For Namu's short time on the waterfront in Seattle, it is amazing all these years later how well he is remembered; Namu made an indelible mark. Many adults over sixty who grew up in Seattle knew the Namu sensation firsthand, and many have home movies of family visits to Griffin's aquarium. Several biologists working with orcas today have told me that when they saw Namu when they were kids, they were inspired to go into marine mammal biology as adults. Namu was quite the celebrity orca, and his skull and teeth are kept for posterity at the Burke Museum of Natural History and Culture in Seattle. One day I went to see them.

Mammalogy collections manager Jeff Bradley had Namu's skull waiting for me on a table and his teeth out of their box and arrayed on a Styrofoam pad. Each tooth was numbered in tiny handwritten script for its place in the orca's mouth. I picked up a tooth and was astonished at its size and weight. It was as big around at the base as a fat carrot, as long as my palm, and dense, hard, and cool to the touch as marble. Each tooth came to a ripping-sharp point, and deep sockets in the jaws showed how the teeth are arrayed to interlock like vice grips, to tear apart prey. The bone of the skull was creamy ivory in color, its shape smooth and rounded, made for moving through the wild water Namu was born to. The skull, big as a coffee table with its bony case to protect the brain, spoke louder than any words in a scientific paper about the intelligence of this animal. No wonder that after Namu's death Griffin, blaming himself, was inconsolable.

I asked him to tell me what it felt like to lose Namu, and the man who had happily showed me what he calls his memory room—full of boxes of clippings, Namu memorabilia, and home movies of orca captures and frolicking with Namu—suddenly clamped shut. "No," he snapped, turning away.

Even after Namu's death, the world was, because of Griffin, orca crazy. The "vermin" shot on sight just a decade earlier was now the darling of the public. Orders for live orcas were streaming in to Griffin's business from around the world. Everybody wanted to see a captive performing orca. Namu had become Namu, Inc. Griffin went back to catching orcas.

Death at Penn Cove

It was August 1970—the peak year for orca captures—when Griffin found himself with far more orcas behind nets than he had ever dealt with or intended to catch. By chance, he had a super pod on his hands: a gathering of ninety to a hundred orcas—possibly the entire southern resident population—surging and leaping behind nets at Whidbey Island's Penn Cove. Alarmed, Griffin knew he had more captives than he could safely handle.

With no regulations whatsoever to restrain him, Griffin could have sold every last orca. But most "croppers," as orca catchers called themselves, still sought—within their limited and often incorrect understanding at that time of orca biology—to manage the live-capture orca fishery competently. Griffin ordered most of the orcas freed, angering his late business partner, Don Goldsberry, as well as the fishermen who had made the initial set of the net and had been promised $2,000 for every orca captured.

That left some forty orcas still captive, and Griffin set about the work of sorting the ones to keep from the ones to set free. Even after their release, the orcas he set loose stayed close by. They would not leave family members still trapped on the other side of the nets.

Up until then the captures were covered in the media, including *The Seattle Times*, like sporting events. But this capture, during the height of the summer tourist season on Whidbey Island, was different. Up close and personal, the capture was harrowing to those who had never seen or heard such a thing before. Although the capture was legal, witnesses interviewed today say that, to them, it still felt terribly wrong.

Terry Newby was on hand to help with the Penn Cove capture that began August 7, 1970, and remembers it well. Captors working in skiffs herded the orcas with firecrackers, then encircled them with nets, separating the young from their parents. "They were frantic. The cows and calves were separated, and there was a lot of crying when they took out the little guys, a lot of tragic cries at night, you could hear them across the water. A lot of spy-hopping, a lot of standing up and looking around; they were continually searching for a way out. It was pretty sad," Newby told me.

"It affected me to the point that I severed my relationship with Don and Ted. To this day, I am still

TOP: The orca who became known as Shamu is readied for her December 20, 1965, flight from Sea-Tac Airport in a cargo plane to San Diego and transfer to SeaWorld. Don Goldsberry, assistant director of the Seattle Marine Aquarium, stands by as she is moved. *(Vic Condiotty/ The Seattle Times)*

BOTTOM: Ted Griffin's aquarium at Pier 56 was the holding and transfer point for killer whales sold and shipped all over the nation and the world. Here, four killer whales await transfer August 16, 1970. *(Ron De Rosa/*The Seattle Times*)*

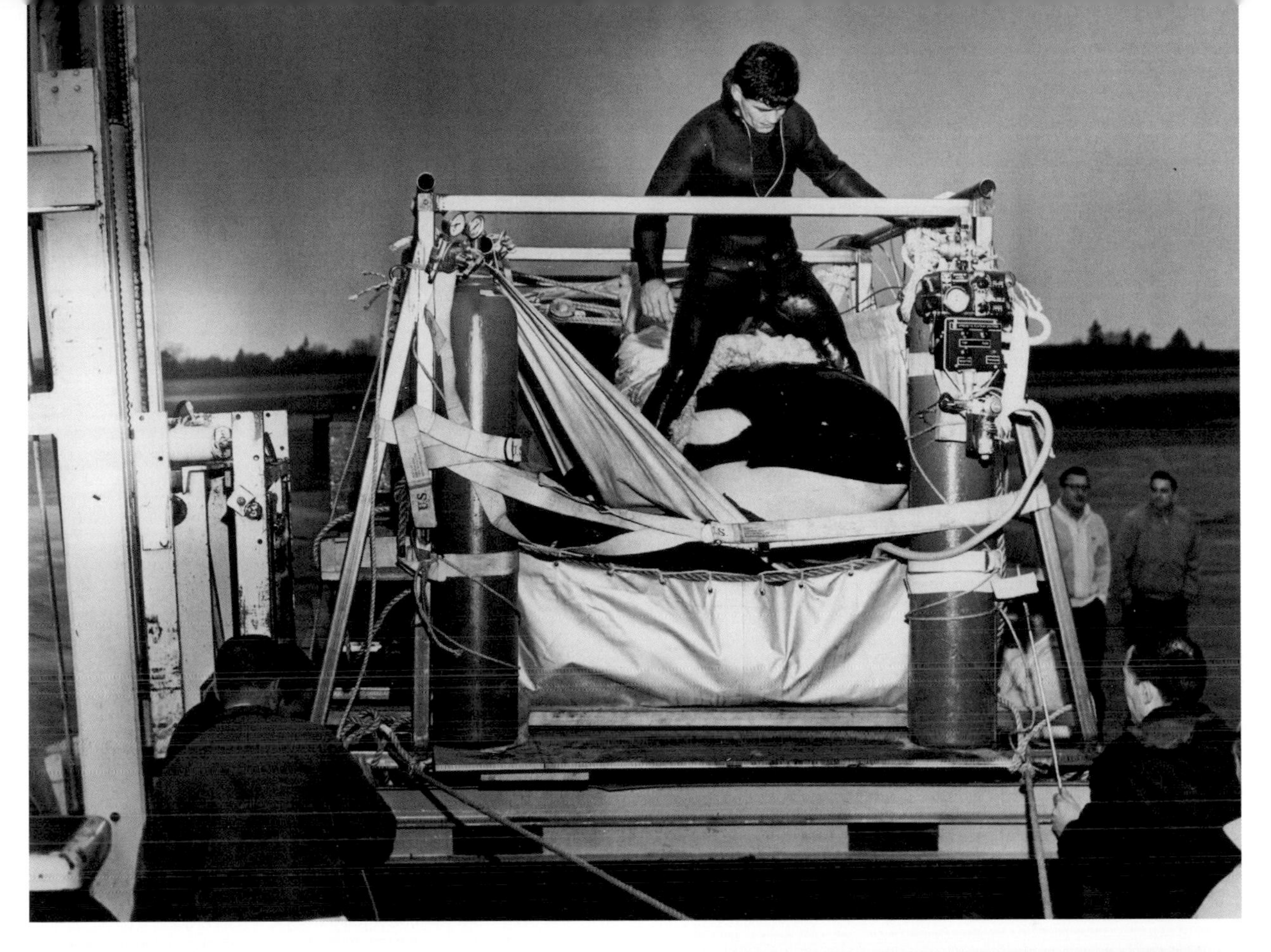

CAPTURED, KILLED: A BRUTAL ERA TAKES ITS TOLL

Between 1962 and 1976, the Pacific Northwest was the world's only source of orcas for aquariums. In those years, about 270 orcas were captured, some more than once. Of those whales, at least 12 died during capture and more than 50 were kept for display. Of those, all the southern residents taken have since died but one.

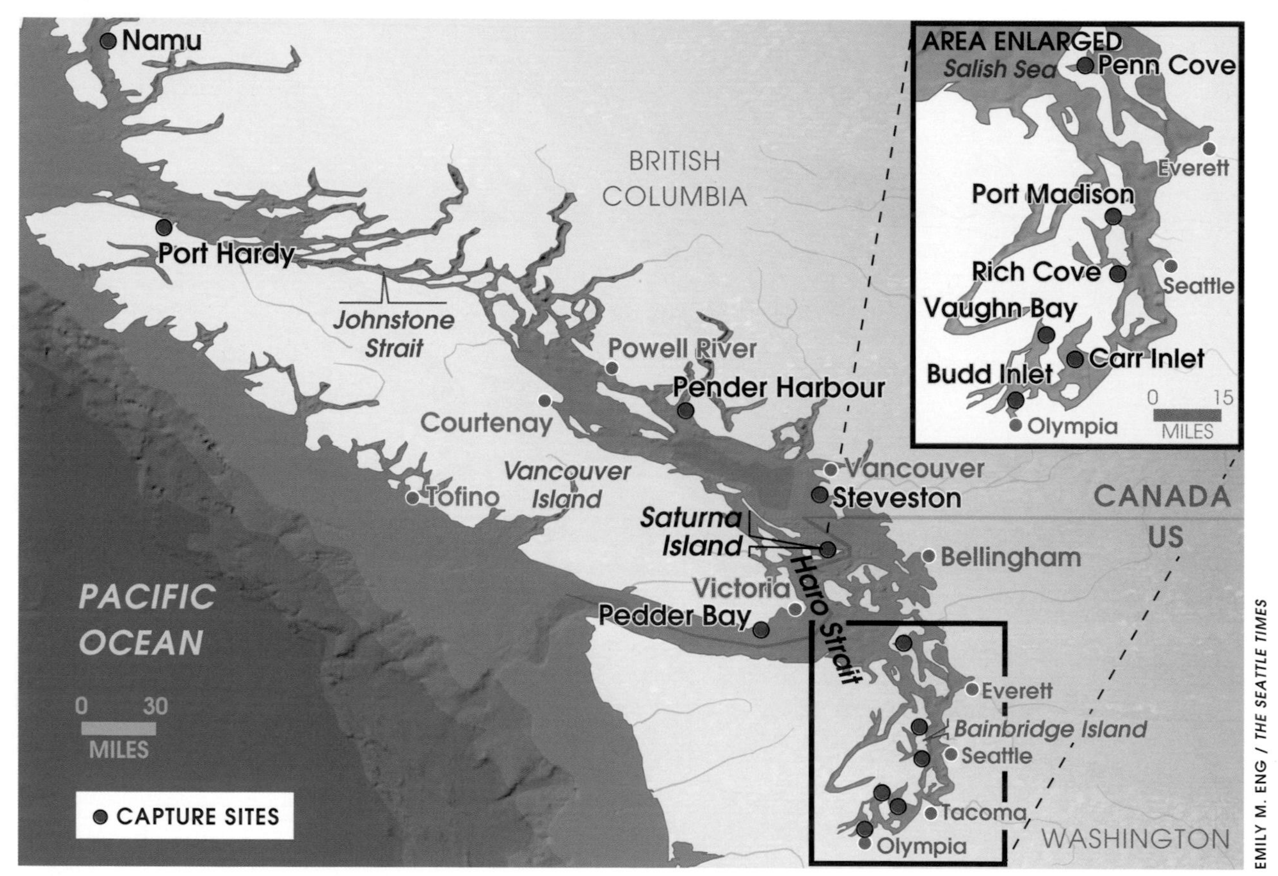

Sources: Jason M. Colby, Orca: How We Came to Know and Love the Ocean's Greatest Predator *(Oxford, 2018); Erich Hoyt,* The Whale Called Killer *(Camden House, 1981); Esri; Natural Earth*

troubled by how the whales felt, seeing and hearing all that."

Lolita, the last surviving captive southern resident, was taken that day. She was one of the orcas Griffin personally selected as a keeper. I've seen his handwritten pencil note recording her catch and her size. With the hunt over, it was Newby's job to keep Lolita calm on the flatbed truck as she was taken to Seattle for shipment to the Miami Seaquarium, where she still lives today. "I was touching her, rubbing her; she was just a beautiful animal and very scared, of course," Newby said.

Marine-mammal captures became big business. Here, on October 25, 1973, two orcas and two belugas were hoisted from their tanks at Seattle's Pier 56, trucked to Boeing Field, and flown to San Diego's SeaWorld. *(Johnny Closs/* The Seattle Times*)*

"She would follow you with her eye. I will never forget that."

It got worse. Someone had cut the nets, intending to free the orcas, but four baby orcas became entangled, drowned, and died. "It's not their fault, you could say," Newby said of the captors, reflecting on the deaths of the drowned orcas after somebody cut the nets. "But it was. Because they already had them in nets." Goldsberry decided to try to cover it up. "He said, 'I am going to take care of it so they will never be found,'" Newby said.

He went with Goldsberry on the ride down to Seattle to get anchors and chains to sink the baby orcas, first slitting their bellies open to let the gasses escape as they rotted, so the carcasses would not float. In all, five orcas were killed in the Penn Cove capture, the four drowned calves and an adult killed later when she, too, tangled in a net and drowned.

The spectacle of the capture and cries of the orcas had already changed the minds of people who witnessed it that day—and many more who read about it in newspapers and watched footage of it on TV. The Progressive Animal Welfare Society picketed Griffin's aquarium, with opponents carrying signs that read "Stop the Killing" and "End the Profits and Greed." That was *before* one of the anchored corpses washed ashore, then eventually three more, dragged from the bottom by a fisherman. The public regard in which Griffin and his capture team were held took a dark turn.

Here were mutilated bodies of baby orcas, cut open and sunk, with anchors tied to their tails in an attempt at concealment. It was beyond gruesome; these were not natural deaths. Once thought of as dashing heroes so brave to take on these fearsome creatures, the captors were seen in a whole new light—particularly as the orcas revealed themselves during the captures to be wholly nonviolent, even when their babies were being taken.

Regulating an Orca Live-Capture Fishery

In 1971 in the public outcry following the Penn Cove capture, the State of Washington set itself up in the orca-hunting business, placing limits on size and charging permit fees of $1,000 per orca. That same year, *The Seattle Times* published its first editorial calling for an end to orca captures.

I think about orca L25. As of 2020, she was the oldest of all the southern resident orcas, born probably around 1928. She must have been captured multiple times and either escaped or was set loose because she was too big. Was she one of the parents frantically watching as their young were taken away? With virtually the entire population of southern residents there that day, surely L25 was too. Is Lolita her daughter?

There is no doubt that Lolita still has living relatives today in Puget Sound, including L25, who is old enough to have witnessed what happened in Penn Cove. On August 20, 1971, what was then called the Washington State Game Department permitted another capture in which fifteen members of L pod were captured, once again in Penn Cove. Three orcas were taken from that group, all of them shipped to SeaWorld in San Diego. Not one is alive today.

As for Griffin, he quit the orca-catching business for good in 1972, selling it to Goldsberry, who then resold the company to SeaWorld. Griffin informally changed his name and moved to Eastern Washington, where he sought obscurity, working as a day laborer for as little as two dollars an hour. His marriage and his finances were in ruins. It was partly heartbreak over Namu but also disgust with what had evolved from freewheeling orca roundups to a regulated business. He didn't like the government telling him what he could do and how he could do it. A fanatical libertarian, he felt his individualism and personal vision and agency were being violated by what the state-regulated orca-catching business had become. The hero exploring a new frontier who had been given the key to the city of Seattle for his derring-do had become a pariah in the public's eye.

There was irony here. Griffin, more than any single person, had ushered orcas into their eminence, bringing them in captivity up close to people and almost single-handedly transforming the public's understanding of them. Griffin said he did not foresee what that new understanding of orcas would mean for him. "I had no idea at the time that would start a thought pattern that would bring my career to an end," he told me.

The federal government enacted the Marine Mammal Protection Act in 1972, ending orca captures in the US and rendering capture permits issued by the State of Washington null and void. But SeaWorld complained it had invested heavily in new marine parks and would be bankrupted if it couldn't get the top-paying attraction: live performing orcas. The National Marine Fisheries Service cut SeaWorld a break on Valentine's Day 1973, granting a hardship exemption without a public hearing.

The trouble for captors was that the orcas were getting harder to catch. They had grown wise to the sound of Goldsberry's boats and tactics, including being chased by helicopter and speedboat and herded with firecrackers. Some of the orcas, by then caught multiple times, had learned to roll their bodies, laying down their dorsal fins on the water's surface, making them harder to spot. The pods would also split up, with parents sending the young in another direction and deploying themselves as decoys.

The Last Orca Hunt in America

After getting skunked by the orcas for years, Goldsberry in 1976 launched yet another hunt for SeaWorld under NOAA's economic hardship exemption, this time in Budd Inlet, near the Washington State capital of Olympia in southern Puget Sound. It would prove historic.

Ralph Munro would go on to be elected five times as secretary of state for Washington. But the day Goldsberry set out for the Budd Inlet hunt, Munro was a staff aide to then governor Dan Evans. Munro was out sailing when he encountered orcas zooming

TOP: The May 1976 orca capture in Budd Inlet at Olympia would prove historic. *(Washington State Archives)*

BOTTOM LEFT: It took a fight by multiple state officials and a court battle to free the whales captured in Budd Inlet and make it the last orca capture ever in America. *(Washington State Archives)*

BOTTOM RIGHT: Loaded in slings and hoisted by a crane, these orcas captured in Budd Inlet were headed for captivity until opponents, including Washington state officials, intervened. *(Washington State Archives)*

across the water and quickly realized the orcas were being chased. He saw Goldsberry and his crew lighting seal bombs—underwater firecrackers—one after another, throwing them at the orcas, and pursuing the terrified animals in speedboats. The orcas were frantic, surfacing and rolling and swimming in circles as the captors ran at them with their boats and kept throwing seal bombs to corral and then contain them in a net. They were trapped.

"It was gruesome," Munro said, grimacing at the memory forty-two years later. "As they closed the net, there was a guy on the back of the boat with a torch, and he was lighting and dropping these explosives as fast as he could light them—*boom, boom, boom*. The orcas were screaming . . . I can still hear them, screaming back and forth. They had parts of the pod inside the net and parts of the pod outside the net. It was just panic, totally disgusting. Sickening."

Munro was desperate to intervene. He reached out to the press, and the next day the hunt was front-page news around the region. He next enlisted the help of the state attorney general, Slade Gorton, later a US senator who served two terms. Gorton mustered a legal attack on SeaWorld. In a news conference, Governor Evans, who also would go on to serve in the US Senate, announced his opposition to the hunt.

Within hours, the state obtained a federal order prohibiting SeaWorld from moving the orcas, which were by then circling in nets surrounded by more than a hundred opponents in kayaks and canoes and on the beach. Munro personally served the order on Goldsberry, approaching his boat in the middle of a dark, rainy night in the company of the biggest game warden he could find.

SeaWorld went to work the next morning to fight back. Legal wrangling to the US Ninth Circuit Court of Appeals and back landed the controversy once more in federal court in Seattle, where demonstrators lined up on the courthouse steps and down the street, demanding a stop to the hunts. "Save the whales," Munro said. "Nobody had ever done anything like it before."

Within a few days, three of the six captured orcas escaped. A fourth was too large to keep and was let go, and two more ultimately were released back to the wild. SeaWorld, in its settlement agreement, vowed never again to hunt orcas in Washington waters. "We won," Munro said. "It was the last [orca] whale hunt ever in America."

Lolita

But for Lolita, the capture era still goes on, despite multiple attempts to free her by animal rights activists, orca advocates, marine mammal experts, orca scientists, and political leaders, including the

LEFT: Former Washington secretary of state Ralph Munro smiles in July 2018 as he remembers winning the battle to end orca captures in Washington waters with the help of the late Slade Gorton, then state attorney general, and Governor Dan Evans. *(Steve Ringman/*The Seattle Times*)*

OPPOSITE: Once celebrated as heroes, captors lost popularity with the public as more people witnessed orca captures in local waters. Demonstrators in May 1970 picket outside Ted Griffin's Seattle Marine Aquarium. *(Richard S. Heyza/*The Seattle Times*)*

late Mike Lowry, a former governor of Washington and member of Congress. Now Lummi Nation tribal members and political leaders are working to free Lolita. In 2020, she began her fiftieth year in captivity.

Lummi tribal leaders have proposed retiring Lolita from performances at the Seaquarium in Miami and bringing her back home, to care for her in a netted-off cove in her home waters, where at least she can once again hear the voices of her family. Perhaps, in time, she could even recover enough to once more go free. However, the Seaquarium has repeatedly refused to release her, saying she is better off at the Seaquarium than at home, where her relatives struggle to survive.

"There is no room for further debate on what is best for Lolita," said a statement from the Seaquarium provided to *The Seattle Times* in July 2019. "Instead of focusing on a perilous move that could endanger the life of Lolita, a fifty-four-year old whale, the attention of those concerned should be on the plight of the critically endangered killer whales of Puget Sound, near the home of the Lummi Nation. Each day during Lolita's presentation at Miami Seaquarium, we focus on the plight of these endangered animals and how we can help them. We all need to come together to understand the threats that are facing the Puget Sound ecosystem and focus our attention on saving the southern resident killer whale population."

For the Lummi Nation, this fight is about much more than one orca. It is about defending home, the Salish Sea. "There was a time not very long ago that we shared the Salish Sea with the orcas, the salmon, and the Salish Sea was rich with the noise of the natural world, and we were a part of that. We are

LEFT: Lolita is loaded for transport by truck. "She would follow you with her eye," says Terry Newby, who tried to soothe her on the ride. *(Dr. Terrell C. Newby)*

OPPOSITE, LEFT: Lolita, also called Tokitae, has been held in captivity at the Miami Seaquarium since 1970. Of more than fifty whales taken, mostly from the critically endangered southern resident population, she is the only southern resident still alive. *(Harley Soltes/*The Seattle Times*)*

OPPOSITE, RIGHT: Still wild, this young female orca is wedged between two adults who are perhaps trying to hide or protect her during the capture at Penn Cove in August 1970. She was named Lolita by her captors at the Miami Seaquarium, where she was still performing in 2020. *(Courtesy of Wallie V. Funk photographs, Center for Pacific Northwest Studies, Western Libraries Heritage Resources, Western Washington University)*

the indigenous people of this land and they know our songs," said Jay Julius, former chairman of the Lummi Nation and a lifelong fisherman. "They are older than us. They were here before us. She represents the true history of what our people have gone through.

"This isn't only about saving a whale; it is about reflecting on our past and what we allowed to take place. Who the hell are we? What have we done? Everything is connected," Julius said. Lolita's loss from the Lummi community and her living relatives still in Puget Sound connects to a larger sundering and loss. "She was pimped out," Julius said. "Puget Sound has been pimped out."

Lolita is approximately 20 feet long and weighs about 7,500 pounds. According to reports by four expert witnesses unsealed in a court case filed in an attempt at her release, she is captive in the smallest tank used for an orca anywhere in North America, sharing it with two Pacific white-sided dolphins that bite and harass her.

She also daily endures an extreme violation of her natural behavior. Wild southern resident orcas travel more than 70 miles a day in their seasonal rounds hunting salmon all along the Pacific coast from central California to Southeast Alaska and throughout the Salish Sea. Their deep, lung-purging exhalations trigger an intuitive kinship: orcas are not fish; they are mammals, like us. And like us, they have lively minds and emotional needs.

"The misconception about whales and dolphins and taking them into captivity is that they are so smart they will figure it out. I think it is the opposite," said Lori Marino, a neuroscientist by training and an expert on the structure and function of the orca brain. In addition to her scientific work, Marino also leads the Whale Sanctuary Project, a nonprofit seeking to create a place for retired performing cetaceans to live out their lives.

Marino documented the harmful effects of captivity on orcas in a paper published in 2019 in the *Journal of Veterinary Behavior.* "They are so intelligent, and their needs are so great, they cannot even get close to having their needs met in an amusement park. Rather than being a buffer, their intelligence is a risk. It makes it more difficult to adapt to that kind of environment. They become more vulnerable, more bored."

Yet orcas are the third most common species of cetaceans kept in aquariums and marine theme parks. Around the world, sixty-three orcas were held in tanks as of 2019. And while live captures have ended in the US, they still continue in Russian waters to supply a growing marine-park industry in China and elsewhere.

Lolita hasn't seen or heard another orca since the death of her tank mate Hugo in 1980. The captive orca had been known to vocalize with L pod calls in her tank. Keith Morrison, a correspondent for NBC's *Dateline* based in Southern California, remembers well what happened when, just out of curiosity, he played L pod calls for Lolita when he visited her on a reporting trip in 1997.

In footage of the event, she surged across her tank and repeatedly zoomed her head out of the water, to put her ear as close to the speaker as possible. "She was very interested. It was like you heard some relative's voice and you recognized it, and you just wanted to hear it more, hear it more,

hear it more," Morrison told me in a phone interview. "It was like she was leaning in. They have very good hearing; she could have heard it from a long way away, but she kind of squished over to our side of the pool and listened intently. She didn't seem interested in leaving, as long as we were playing the tape."

He was shocked at the size of her enclosure. "It would be like someone put you in a bathroom and closes the door and you can't come out the rest of your life. You would be alive, but what kind of life would it be? It was that tiny pool, that was what was so shocking. You think there is some big expansive area for her to go back to after the show is over. That was it; it is terribly small and cramped. Having heard how they travel 70 miles a day and all the activities they engage in, to see this thing restricted to just swishing back and forth in this small and not very deep pool, it was just sad. Like seeing someone in solitary confinement.

"It's been a long time, but the one thing that stands out for me is the reaction to the tape. The pool was tiny; she seemed very cramped in it. She did the show, and we watched it, but the far more interesting moment was her connecting with that sound and with us. She would come up and check you out; she wanted to hang out for a bit." He did not remember whether she made any sounds back.

In a 2018 federal appeals court case, a three-judge panel ruled that it could not identify a threat of serious harm to Lolita, then estimated to be fifty-one years old, that could trigger a federal animal-welfare-law violation. The court also was not convinced there was a realistic way to return her to the wild without being harmed.

Yet the reports by four expert witnesses unsealed in the unsuccessful lawsuit to free Lolita painted a gruesome picture of her day-to-day life. John Hargrove, former senior trainer at SeaWorld, stated in his testimony in February 2016 that "Lolita's tank at the Miami Seaquarium is without question the smallest and most barren I have ever seen an orca forced to live in. The tank measures only 20 feet deep at its deepest point . . . and the tank level is often dropped. . . . It is absolutely unthinkable that an orca measuring at least 20 feet long and weighing over 7,000 pounds lives in this outdated and inadequate facility, and without any same-species contact. It is difficult to believe that this facility still exists in the United States, where even many dolphin facilities from decades past that were essentially this size have long since been closed."

The dolphins she lived with bit and harassed her to the point that she was administered painkillers and antibiotics to treat open wounds, according to Hargrove's statement. The dolphins even bit her on her tongue more than seventy times, Hargrove stated, "an injury I have never seen on an orca." He concluded she was neither healthy nor thriving and exhibited stereotypic bored behavior patterns from living in such a small and sterile confined space, including floating motionless for hours at a time.

In her expert-witness report filed in February 2016, New Zealand–based orca researcher Ingrid Visser also found the tank too small and asserted that Lolita is severely restricted in what would be her natural movements. She is unable to dive at all, Visser noted, and deprived of swimming or hanging vertically without hitting the bottom of the tank. The tank is divided by a large concrete island used for a stage in her performances, further restricting

At a May 2018 interfaith gathering in Bellingham, Washington, supporters lay hands on an orca totem pole carved at the Lummi Nation to raise awareness of the killer whales' plight in Puget Sound and to press for Tokitae's release. *(Steve Ringman/*The Seattle Times*)*

her space. It also frustrates her biological need to swim long distances. Tokitae, as Lolita also is called, would have to lap her pool 24 times to log just one mile and 2,400 laps to rack up the miles that would be typical for her in a day in the wild, Visser calculated.

Mostly, she is so bored she spends an inordinate amount of her time logging at the surface or lying on the bottom of her tank, staring at the wall at an intake valve and its flow. "From my personal observations and from my review of the evidence, it is clear that Tokitae is generally lethargic, that she consistently exhibits an apathetic behavioral state in which her stupor is displayed to the point of her being basically catatonic for much of the time . . . I was shocked and horrified at the abysmal conditions Tokitae is held in." Two more experts filed similar reports.

For the Lummi people, this restriction of Lolita's natural cultural and emotional range and even her physical confinement is all too familiar. Beginning in the 1860s in the US and Canada, the children of the Lummi Nation and other Native communities were taken from their families and sent to government-run boarding schools. Not really schools at all, these were reeducation camps meant to "kill the Indian and save the man," in the words of Richard Henry Pratt, the late nineteenth-century former army officer and founder of the now infamous Carlisle Indian Industrial School in Pennsylvania. Removed to places far from their families, the children's hair was cut, their clothes were burned, and their names were changed. They were forbidden to speak their language and punished if they did. Many were abused physically and sexually. Many died.

"We understand what happed to her," the late Tsi'li'xw Bill James, the Lummi hereditary chief, said of Lolita. "We can relate to her captivity because of the things that happened to our people. Our young people were taken away from us. Our families' cultural values were taken away from us, just as hers were taken away. That's how I relate to her, with her captivity, and the government schools taking away our children, and the trauma, taking away our history, our language, our culture, the breaking up of our families.

"I can feel her heart. I can feel her lonesomeness. And how it is to be taken away from home and family. Because I have heard all the stories of our ancestors of being taken away . . . I know what happened to them, and the hurt they carry, and the historical trauma that has hurt our people today with the loss of our tongue and the loss of our cul-

LEFT: Sul ka dub (Freddie Lane), a former member of the Lummi tribal council, has joined with others from the Lummi Nation and beyond to work for Tokitae's release from captivity. Here he is painted for spiritual protection during a June 2019 ceremony at the waters where she was captured. *(Nancy Bleck)*

OPPOSITE: In the wild, orcas are always on the move, diving, spy-hopping, breaching, and traveling as much as 70 miles a day—natural behavior impossible in captivity. *(Dave Ellifrit/Center for Whale Research; taken under NMFS Permit 15569 and DFO SARA Permit 388)*

ture. We are trying hard to bring it back. She still sings her family's songs. One of the reasons she is still alive is she still knows who she is."

Just like the Lummi people, whose identity also has never been extinguished. Even when forced to cede most of their ancestral lands to the US government in an act of survival, their ancestors insisted on reserving their fishing rights in their home waters, to preserve their way of life and provide for their people. A core of the Lummi's old territory remains in their possession, too, alongside the waters of the *qwel lhol mech ten*, the people that live under the sea.

A Fight About More than One Whale

On a drizzly morning in May 2018, the hereditary chief and Sul ka dub (Freddie Lane), a former Lummi tribal councilman, bounced in a pickup truck over a sand spit that emerges on the first daylight low tide every spring. It was time to go to Portage Island, an old home ground of the Lummi people uninhabited today, but still used for ceremony—and clamming. Lane set to work digging a pit in the sand and gathering beach stones, while James collected driftwood for a fire. As the flames built and sweet smoke rose, more tribal members arrived and headed to the tide flats to dig clams. There was laughter, storytelling, and soon feasting. Shucked, cleaned, and roasted by the fire on hand-cut ironwood sticks, the clams were crunchy and smoky on the outside, sweet, creamy, and succulent within.

Lane heaped rocks on a second fire over the sand pit, piling them high and baking them until the rocks were suffused with a golden orange glow and so hot that they started to split. Only then did Lane pour on buckets of clams, then heap verdant green seaweed, fresh, wet, and briny, on top. "My

Indigenous people of the Coast Salish territories in the United States and Canada hold orcas in high regard and consider them family members. The Namgis understand they are their reincarnated chiefs. *(Mark Malleson/Center for Whale Research; taken under NMFS Permit 15569-01 and DFO SARA Permit 388)*

secret recipe," Lane said, as plumes of steam carried the fragrance of clams on the fresh sea breeze.

Soon elders and children were sharing in the feast. Seeing a young tribal member taking a particular interest in the preparation, James gifted him his roasting sticks, all but the one carved with the name of his mother, who had passed the sticks on to him. As I listened to the visiting going on all around me, to the *shush-shush* of the tide, to the sound of rocks being piled on the fire, I thought, *How long have these sounds, scents, tastes, skills, and stories persisted, in just this way, in just this place?* The soft entreaties of the gulls, the heron stalking the shallows, the chitter of eagles, the gossiping clams, squirting in the clean dark-gray sand. How long?

The clambake was put on to honor what would have been James's late mother's ninety-fifth birthday. Even in simple feasts like this day's gathering, culture is carried on. James said he is encouraged by the revival underway in Coast Salish communities, where tribes, irrespective of the US-Canada border, share a cultural continuum and history that have persisted across hundreds of miles and thousands of years.

But the state of the natural world and in particular the battle for survival by the southern resident orcas saddened James. It was a more than abstract grief; it was a personal anguish and a cultural assault. The Lummi's relationship with animals on sea and land and forces of nature that define, inform, and guide their culture is imperiled. Orcas, salmon, clean water, and even the very rhythm of the seasons are under threat as the same non-native expansion that so disrupted and displaced their ancestors continues today to alter the land, the waters, even the climate. The Lummi people's ancestral values also have been violated. And in the captivity of Lolita, that violation continues.

The Miami Seaquarium has never faced a challenge quite like the Lummi tribal members' quest to free Lolita. So far, the Seaquarium has, in its public statements, acknowledged the tribe but waved away any move to free Lolita. "Miami Seaquarium would like to reiterate our respect for the people of the Lummi Nation and applaud their recent efforts to bring attention to the challenges facing the critically endangered killer whales of the Pacific Northwest," Eric Eimstad, general manager of the Seaquarium, wrote in a prepared statement in June 2019. "For almost five decades we have provided and cared for Lolita, and we will not now allow her life to be treated as an experiment. We will not jeopardize her health by considering any move from her home here in Miami."

In addition to the park's determination to keep Lolita, her release is bureaucratically complicated. NOAA would have to approve Lolita's release, considering both her safety and that of orcas in the wild, which are struggling to survive.

Historian Jason Colby said it is too easy today for people to blame the decline of the southern residents on orca catchers like Ted Griffin. Colby often reminds me that while the southern residents were hard hit by the captures, what is hurting them now is more insidious—and, unlike the capture era, still goes on. It is the loss of healthy waters that produce the chinook salmon southern resident orcas need to survive. The pollutants that poison their bodies, their food, and even orca mothers' milk. The racket people make with boats and ships and underwater construction, making it hard for orcas to hear well enough to hunt.

If Lolita were freed, she would come back to home waters both degraded and depleted. In what used to be a place of wild splendor and abundance, her extended family is hungry in hostile waters.

CHAPTER THREE

Hunger

The crew of the *Bell M. Shimada* hauled in the net, long as a football field and teeming with life. Scientists aboard this NOAA research vessel for a weeklong June research trip off the coast of Washington crowded in for a look.

As the quest to understand why Puget Sound's orcas are in decline continues, a key area of investigation is the primal necessity of regularly available, adequate,

A sea lion rips into a chum salmon swimming home to spawn in the Nisqually River in January. A resurgence of seals, sea lions, and northern resident orcas is a victory for conservation. But the resurgence of marine mammals is taking a bite out of the southern resident killer whales' food supply. *(Steve Ringman/* The Seattle Times*)*

Lack of access to enough chinook and other salmon where and when they need it intensifies all the other problems Puget Sound's killer whales face. Scientists returned this fish, caught in their survey net, to the sea. *(Steve Ringman/*The Seattle Times*)*

quality salmon. Aboard the *Shimada*, scientists were hoping to learn what factors are influencing the survival of chinook, the orcas' primary food.

To document the ocean conditions for migrating juvenile salmon, they dragged a net through the surface of the water and analyzed the catch. They were interested in the array of species present and, especially, the abundance and condition of juvenile salmon. The fish would be bagged and frozen for later lab analysis to determine everything from where the salmon were born to their condition and rate of growth, and even what was in their stomach.

The net came on deck with the groan of an overhead winch, and the crew worked quickly, sorting the catch with their hands. There were wonders: wolf eels, twining their writhing, foot-long bodies and snapping their toothy jaws; a spade-shaped starry flounder, flat as could be; shining bright silver smelt, long and exquisitely slender; a juvenile chum with its overlapping, perfect tiny scales, gleaming with silver and gold and green highlights.

The beauty of fish is underrated, even unknown. What color is a fish? It depends on the light and its angle, I discovered, and how newly pulled the fish is from the sea. Who knew a sardine is flaked with gold and has an aquamarine back and a silver belly bright as starlight? I held a juvenile steelhead in my hand, amazed at the depth of its indigo blue and malachite green colors and the iridescence of its scales. Bronze kelp was calcified with the white rings of barnacle scars.

Dozens of small squid, no longer than an index finger, were translucent and fragile, with a pale, stippled coloration in lavender, rose, gray, and black. Bearded with a delicate fringe of gossamer tentacles at one end, the squid looked all-knowing, with huge eyes gleaming like gold foil, centered with depthless dark pupils.

Entire plastic totes were filled with jiggling jellyfish, piled in blobs that oozed out of the totes and hit the deck with a distinct wet splat despite the crew's best efforts to corral them. The catch included adult chinook, too, making a rumpus in the net, their fins poking through the webbing, their scales glittering like sequins on the deck and on our foul-weather gear. Anything bigger than a juvenile was quickly hustled back, hulking and gasping, into the sea.

Herring may have been the loveliest of all, blue violet and teal green above, silver below, a bicoloration made for hiding but also a wonder to behold. Each fish, each life, still sparkled fresh from the sea. No eye was yet dulled, color dimmed, or shine diminished.

Weirdly, among the Washington native fishes, there also were pompano, tropical fish with pretty pink highlights, iridescent as a soap bubble, that were not supposed to be here at all. And a lot of those jellyfish by the toteful weren't the right species for this part of the ocean, either, signaling a shift in that community too. None of these anomalies were as extreme as in 2018, I was told. But things were not quite right, or normal either, whatever that even means anymore.

What the scientists see on this survey, conducted since 1998, has taken on new importance as oceans warm in the era of climate change. Decade-long cycles of more and less productive ocean conditions for salmon and other sea life are breaking down. The cycles of change are quicker. Novel conditions in the Pacific are the new normal.

"It used to be up, or down. Now it is sideways," said physiological ecologist Brian Beckman, of NOAA's science center in Seattle, as we talked on the bridge of the ship. It was a place well suited to taking the long view, with its sweeping aspect to the horizon, across a gray sea meeting a gray sky. Only the slightest color shift at the horizon and the water's ceaseless motion hinted which was which. The smell of the sea was all around us—on deck, in the lab, in our clothes. A rain shower blew past in the distance, headed toward the lush green land.

Oceans Heating Up

Changing ocean conditions, and marine heat waves in particular, are bad news for endangered orcas that rely on salmon for food. The cold upwelling of nutrient-rich waters sustains salmon while they are at sea, enabling them to grow and fatten. But warm water conditions mean poorer nutrition for salmon. This was something scientists on the *Shimada* were on the lookout for. Nighttime was prime time for towing a net alongside the ship to gather samples of tiny animals called zooplankton to assess the ocean's food supply for salmon. The crews got up twice each night, the ship ablaze with lights, to capture zooplankton migrating upward in the water to feed on plankton, the great green pastures of the sea. This I wanted to see, so I set my alarm to get up with the survey crew.

Rich Osborne arrived on deck with a pair of bongo nets, so-called because of their shape. A lifelong salmon and orca advocate, Osborne was aboard the ship as a volunteer from the University of Washington's Olympic Resources Center, helping out with whatever needed doing—including midnight zooplankton surveys. An assistant from the *Shimada* crew helped him carry the nets to the ship's rail, holding the delicate netting like an attendant with a bridal train. The night mist was swirling

LEFT TO RIGHT, TOP ROW: Pacific herring (*Clupea pallasii*), native, can reach up to 18 inches long. Herring are crucial food for salmon; Salp (*Thetys vagina*), native. This lone specimen was palm-sized. It moves by pumping water through its gelatinous body; Caprellid amphipods (*Caprella* spp.) on kelp, native, about the size of a corn kernel. All are part of the ocean food chain.

MIDDLE ROW: Wolf eel (*Anarrhichthys ocellatus*), native, average 20.7 inches long; Juvenile chinook (*Oncorhynchus tshawytscha*), native. A typical mature chinook is 36 inches long; Sea nettle (*Chrysaora fuscescens*), native, average 6.2 inches long.

BOTTOM ROW: Pacific pompano (*Peprilus simillimus*), nonnative; Pyrosome (*Pyrosoma atlanticum*), nonnative, average 5.1 inches long; California market squid (*Doryteuthis opalescens*), native. *(All photos by Steve Ringman/* The Seattle Times)

in the lights like snow as Osborne gave a thumbs-up to the captain watching from the bridge to indicate he was ready for the tow. Osborne held a protractor aloft to make sure the ship's speed was just right to trail the nets through the water at a 45-degree angle to the current.

After a short tow, the crew lifted the nets and Osborne rinsed the catch off into a sieve. Then Jenny Waddell, research coordinator for the Olympic Coast National Marine Sanctuary, moved the material he had captured into a jar with preservative, for later analysis. I asked if she could save out a sample for me—I wanted to have a look at what was out there. It didn't look like much, this smear of brown goo that she plopped into a mason jar just for me and topped off with seawater. I got up in the middle of the night for this? Surely there was more to see than this bit of sea drool.

But then I held the jar up to the light—and the water came wondrously alive. Here were algae, diatoms, baby crabs, larval fish, pulsing bells of jellyfish, winged snails, the brilliant red bodies of krill—a tiny crustacean that is a favorite food of salmon and gives salmon their signature color.

This one small sample—just a quart canning jar's worth—was a primal soup, seething with small lives zipping, gyrating, twitching, leaping, pulsing, and swirling. One tiny shrimp-like creature furiously sculled through the water with a phalanx of legs tinier than an eyelash. Some of the animals just drifted, but others moved right along. These are the little lives that run the world, feeding the forage fish that feed everything else, including baby salmon.

The sampling crew went back to their staterooms for naps before they would get up again at two in the morning for another round. I took the jar to my berth and got under the covers with a flashlight, to watch the zooplankton just a bit longer without waking my bunkmate. The ship rolled through the long swells as the lively foundation of the food chain carried on in my jar. As I watched tiny creatures glisten in the flashlight beam, I thought about worrisome questions in this primal soup too. Was it fattening enough? Why were juvenile salmon faring so poorly in the ocean, with fewer juveniles caught in the sampling net than hoped? For baby salmon, ocean conditions—specifically, the abundance and nutritional quality of this zooplankton and plankton—are a matter of life or death.

A baby salmon's urgent task is to grow bigger than a bird's beak—fast—or it will never live to feed an orca. Scientists want to see four times as many juvenile fish survive in the sea than currently do. But ocean conditions haven't been that good in decades. Then they got even worse. "When the Blob hit, everything changed," Brian Beckman said.

The Blob

A gigantic mass of warmer-than-normal water off the Pacific coast, which came to be called the Blob, began forming in late 2013. It depleted the ocean's food supply and killed an uncounted multitude of animals, including seabirds and marine mammals. In June 2017, scientists caught so few juvenile chinook they thought there might be holes in the net. Freakish numbers of species such as pyrosomes, a firm tubular plastic-like animal of subtropical seas, covered the decks.

Those most dramatic influences of the Blob had dissipated, said Brian Burke, a supervisory research fish biologist at NOAA's science center and chief scientist on the 2018 ocean survey on the *Shimada*.

Still, in some places where juvenile chinook in past years had been most abundant, very few were caught at all.

So powerful are the effects of ocean conditions that they can swing even abundant runs of salmon into dramatic downturns—or provide a bonanza of spectacular bounty. After decades of little change, more than a million chinook a year came back to the Columbia River system from 2013 through 2015, smashing modern records and capping fifteen years of greatly improved returns since dismal runs through the 1990s. Yet as the full effects of the Blob developed, chinook runs crashed again.

What will ever-more-unpredictable ocean conditions mean for salmon? And the southern resident orcas? "What if the frequency of these events increases, even if they don't get worse?" said Ritchie Graves, chief of the hydropower division for NOAA's Northwest region, about the Blob. "We lost twenty years of investment in improving the status of stocks. We are almost back down to where we were in the bad times of the 1990s."

The Blob turned up the heat on salmon already struggling for survival. Across the Pacific Northwest, 40 percent of chinook runs already are locally extinct, and a large proportion of the rest that remain are threatened or endangered. Yet without more food, the southern resident orcas will be extinct within a hundred years, Rob Williams, chief scientist of the Seattle nonprofit Oceans Initiative, and other colleagues projected in a 2017 paper. What is underway is the decline of twin monarchs: the southern residents and the chinook salmon they principally rely on. Both animals are signatures of the Pacific Northwest, species without which the region would be culturally, ecologically, and, some would argue, spiritually no longer the same. Salmon, particularly chinook, are animals with the power to transform anyone's sense of place. To see chinook on the move in a stream, to catch one, to behold one resplendent in the market or on the table or roasting over a fire, is to know the region's wealth.

King of Kings

I remember well the first time I met a Columbia River spring chinook. I had traveled to southwest Washington from Seattle in a driving rain the night before, the kind of deluge in which every truck swamps the frantically swatting wipers of passenger cars. I'd found the quietest room I could in a highway motel and slept . . . and slept some more. With a start, I awoke recognizing I was dangerously close to being late for the boat I was to meet that morning. I didn't expect anyone to wait for me. And I had an appointment with a fish.

The diesel engine of the skipper's purse seiner was purring when I met him at the dock in Skamokawa, a quiet fishing and logging town along the lower Columbia River. The crew pulled in the mooring lines and we headed off to bigger water. It was calm and gray, the river a mirror, the weather breathlessly still. Waves curled under the bow, and soon we reached the spot for the first set of the net. Before long its plastic floats dipped and bounced: fish.

I remember a controlled pandemonium as I tried to stay out of the way. But most vividly, what happened next was an indelible face-to-face meeting as the skipper lifted and put before me a glorious fat Columbia River spring chinook. I knew these fish were big, but nothing prepared me for the thrashing slab of muscle that is a healthy adult spring king

SALMON STRUGGLE TO SURVIVE IN THE SALISH SEA

Salmon have been in a steady decline since 1980, victims of a range of threats in the waters that include Puget Sound, the Strait of Juan de Fuca, and the Strait of Georgia.

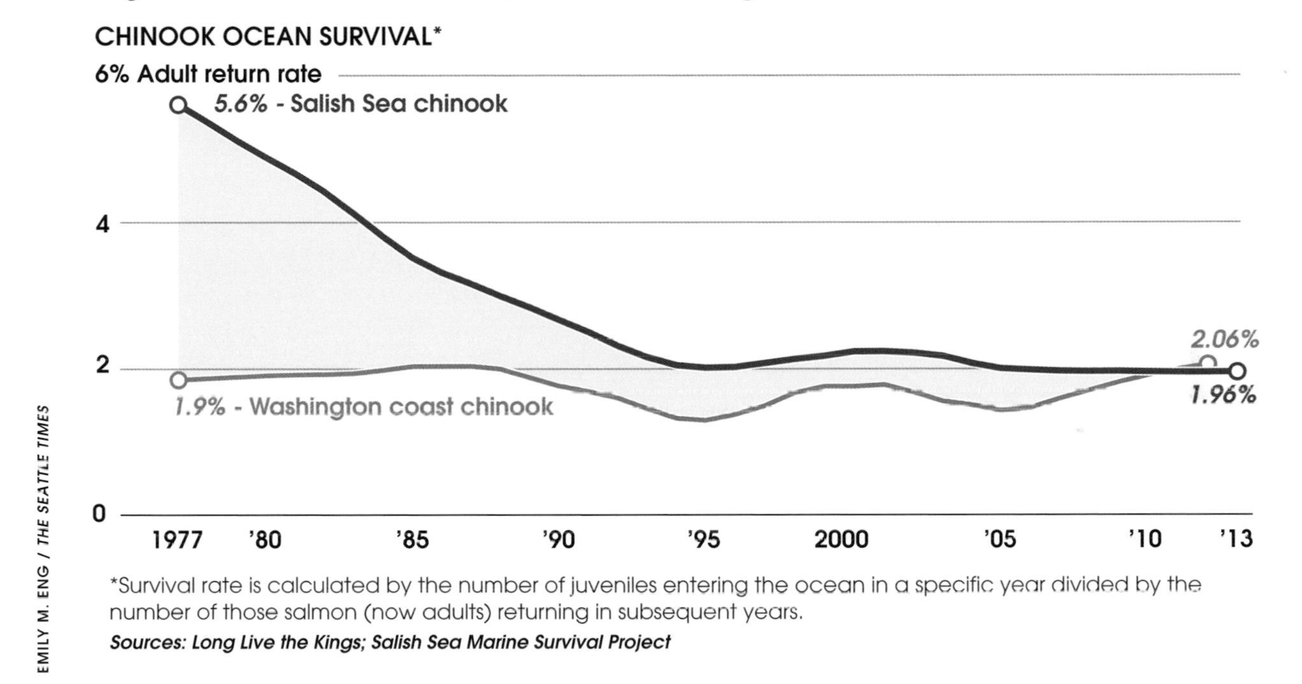

from the Columbia. Its golden eye was wide and staring from all the mysteries it had seen in four years at sea; its body was sleek, meaty, and firm. Its silvery scales were so bright and shining they seemed to make their own light, iridescent lavender, purple, and blue refracted from each disk. I reached out to touch this lustrous fish, feeling the slippery smooth slide from the top of its head to the tip of its nose.

Then it bit me.

I leapt back with a yelp, pushing the blood from the end of my finger in two tiny dots. A wound this was not. Rather, it was an initiation by salmon tooth, a blood bond that would endure. I had met my first chinook, and it had marked me, in more ways than one. No one ever forgets their first big king.

I examined the fish in more detail, noticing it was smooth if I rubbed its scales in one direction, but raspy as a cat's tongue if rubbed in the opposite direction. Those scales had a lay to them, like shingles on a roof. Luminous and supple as a thin suit of armor, they bent and flexed with each movement of the fish.

Salmon scales have rings that scientists read like rings in a tree trunk, enabling them to learn the age of a fish in years spent at sea. A fish has the same number of scales throughout its life (and it

The southern residents eat mostly chinook but in the fall include much more coho in their diet, such as the fine, fat fish this K pod whale has snagged. Later in the year, the southern residents will also take chum, and they occasionally hunt other fish, including steelhead and halibut. *(Candice Emmons/NOAA Fisheries; taken under Permit 16163)*

can grow replacements). The scales grow to fit the fish, becoming bigger and stiffer as the fish puts on size. As an additional protection, the fish is covered with a layer of clear slime that protects it from infection. Slime even helps the fish slip out of tight spots, including the grip of bears and, yes, people. Not an adequately lovely word at all, slime is a thing of beauty on a healthy fish.

The color of the salmon I held comes from pigment in its skin, which changes along with hormonal shifts. Some salmon morph dramatically as they come into sexual maturity, taking on distinct spawning colors and even changing their body shape, growing a humped back and distinctly hooked nose. Chinook arrive silver bright from the sea and gradually darken in color as they migrate to their home gravel in preparation for spawning. All salmon develop bigger teeth at spawning time, which they use like fangs to chase off competitors for mates and to defend their nests.

Each fish is not just another fish. Rather, each chinook population is one of a kind, a glittering jewel in its particular setting with a genetic gift earned through thousands of years of adaptation that makes a population fit for its particular place of origin. It all counts: stream temperature and chemistry, the timing of the spring snowmelt that nudges baby fish downriver to the sea, the homing of each species of salmon to its specific migratory path in the ocean. None of this is random, and all of it is encoded in the genes of the fish, enabling it to thrive in its specific stream and river system.

Once at sea, chinook grow and fatten for as long as seven years—very unusual, these days—but more typically anywhere from two to four years before obeying a mysterious signal and turning toward home to spawn. This perilous and strenuous journey is led by a hardwired instinct to use geophysical cues, from the earth's magnetic field to the chemistry of the water. A young salmon takes up the chemical imprint of its home river. Years later, when it navigates home, the fish can detect the scent of a single drop of its home water in 250 gallons of seawater. It relentlessly follows that smell of home to its spawning gravel.

Even scientists who have studied Pacific salmon their entire career do not entirely understand the wonder of salmon migration to the sea and back. "What makes a fish go where it goes?" said Tom Quinn, fisheries biologist at the University of Washington and a foremost expert on Pacific salmon. "They haven't been there before. And no one they are with has ever been there before. Yet they go more or less to the same place in the broad distribution [in the ocean] of their kind."

Scientists know salmon detect the earth's magnetic field, and they may also have a mapped sense of the ocean, Quinn said. "The fish know where to go, like sea turtles or homing pigeons; it is not just higgledy-piggledy, every one goes every which way. Even if stocks get mixed up at hatcheries, they still figure it out."

Salmon have a sense, when going out to sea, of where they are headed and then, when headed home, of where they started out, no matter the distance or how remote their home waters. "They routinely find their way from the middle of the ocean to some trickle of water in Idaho," Quinn said—a feat the more remarkable given they have no help and a limited view of what lies ahead. Birds can look down and follow older, more experienced birds, Quinn noted. But returning salmon travel without guides more experienced than they are, and the farthest they can see is 10 feet. They can't even look up or down.

Salmon migration over thousands of miles through years in the vast ocean and back imbues the spirit of the Northwest with wildness and wonder, said Quinn, who was raised in New York, "where salmon were on a bagel with cream cheese and bears were in the zoo." Wild salmon are a miracle that still moves him. "The courage of their vaulting at the waterfall, struggling to get home," he said, "there is something heroic about that. We need more heroes."

Salmon *are* incredibly tough. They have radiated since the Pleistocene into every suitable habitat and strayed into new territory as needed because of volcanoes, wildfires, landslides, rockfalls, ice dams, or other calamity. The family Salmonidae goes back one hundred million years and the five species of Pacific salmon were apparently distinct from each other by several million years ago. After the glaciers melted and sea levels stabilized, forests sprouted and grew, creating the conditions for a long period of stability for salmon, lasting about 7,500 years. Over that time, the present populations of salmon coevolved with the lands and waters and natural processes that created the Pacific Northwest.

A Gift of Abundance

According to a Washington Department of Fish and Wildlife report by the late Jeff Cederholm and his coauthors, salmon are a keystone animal, at every life stage from egg to carcass, feeding 137 species: Caddis flies, stone flies, and midges. Bald eagles, river otters, beavers, and, of course, orcas. Bears, herons, coyotes, raccoons, sea lions. Harbor seals. Puffins, terns, ospreys, kingfishers, and dippers. Pacific giant salamanders, ravens, gulls, and sea ducks. Trees and soil and millions of unseen microbes. Even in death, salmon mean wealth writes fisherman and author Roderick Haig-Brown in *A River Never Sleeps*:

> The death of a salmon is a strange and wonderful thing, a great gesture of abundance. Yet the dying salmon are not wasted. A whole natural economy is built on their bodies. Bald eagles wait in the trees, bears hunt in the shallows and along the banks, mink and marten and coons come nightly to the feast. All through the winter mallards and mergansers feed in the eddies, and in freshet time, the herring gulls come in to plunge down on the swifter water and pick up the rotting drift. Caddis larvae and other carnivorous insects crawl over the carcasses that are caught in the bottoms of the pools or against the rocks in the eddies. The stream builds its fertility on this death and readies itself to support a new generation of salmon.

Scientist Ted Gresh of the University of Oregon and his coauthors looked at records of old canneries on the Columbia River to estimate how much nutrition from returning salmon historically fed streams and soils. Nitrogen, phosphorous, and the building blocks of life were carried in the bodies of fish by the thousands of tons, the scientists determined in a January 2011 paper published in the journal *Fisheries*. Salmon also fed the next generation of salmon: in one analysis, scientists found up to 40 percent of the carbon in coho smolts came from nutrients derived from the decaying carcasses of the previous generation. Juvenile salmon eat salmon eggs and feed directly on spawned-out carcasses. They also feast on bugs that have eaten decaying salmon and feed on increased algal growth that is lush with the nutrition gifted by salmon from the sea.

Scientists even planted dead fish in streams to see how salmon carcass supplements affected juvenile coho. They measured salmon growth in an enhanced stream and in one with no added fish. Both the size and the number of juvenile fish were much greater where the biologists had slung the dead salmon. The benefits were lasting: the bigger fish also had better rates of overwinter survival.

Hatcheries Industrialize Salmon

Salmon abundance began to diminish in 1850 with the arrival of settlers. Photos of cannery wharves in Seattle buried in too much fish to can or sell show the greed and waste common at the time. The fish, like the forests, were considered inexhaustible, just there for the taking.

Laws were passed banning fish wheels—brutally efficient mechanical-harvest devices—but that just left more for the overharvest by gillnetters. Lawmakers also prohibited construction of dams without fish passage, though these laws were unenforced. When fish ladders washed out, sometimes within just a few years of being built, the dams stayed and no repairs were made. Logging, mining, dredging, diking, draining, filling, farming—the landscape that sustained the salmon was transformed for settlers' use at an astonishing pace and scale. As the newcomers built their wealth, salmon runs dwindled and streams were impoverished. Then a new idea was born: industrialize the fish too. Grow them, substituting industrial fish hatcheries for the fast-disappearing natural habitat of the salmon.

As hunting depleted waterfowl and elk, no one suggested a duck hatchery or an elk farm. Instead, reserves and national parks were created, where animals could flourish unmolested. But salmon have never had habitat set aside just for them. Instead, the idea was that if there weren't enough salmon, we would rely on human ingenuity to make more. Trays and concrete raceways would replace the forests and streams, rivers, and estuaries needed to raise salmon. We would produce salmon like any other agricultural commodity.

Billions of baby salmon are pumped out of hatcheries in Russia, Japan, Alaska, Washington, Oregon, and California every year. Yet the southern resident orcas continue to go hungry, and many salmon runs are still in decline.

Puget Sound chinook were listed for recovery under the Endangered Species Act in 1999, but after twenty years and hundreds of millions of dollars spent, the ten-year average abundance of natural-origin wild Puget Sound chinook is 28 percent *below* what it was *before* the recovery effort was started. "All the recovery plans, all the effort made no difference," Ron Warren, director of the fish program for the Washington Department of Fish and Wildlife, bluntly said in a 2019 phone interview.

Hatchery practices themselves are helping to bring about the demise of wild salmon. A wild fish loses fitness in one generation of interbreeding with a hatchery stray. Juvenile hatchery fish are bigger than wild fish when first released and will eat smaller fish, including wild salmon. They also compete with wild fish for space and food and can harbor and spread disease.

As adults, hatchery fish lose their size advantage; they are often smaller than wild chinook because

SALMON DECLINING IN ABUNDANCE AND SIZE

Chinook populations up and down the West Coast have slowly been decreasing since the 1980s. Not only are there fewer fish in regional waters but individuals are shrinking in average size and weight, with the older, fatter salmon making up less and less of the population.

CHINOOK ABUNDANCE FROM ALASKA THROUGH CALIFORNIA

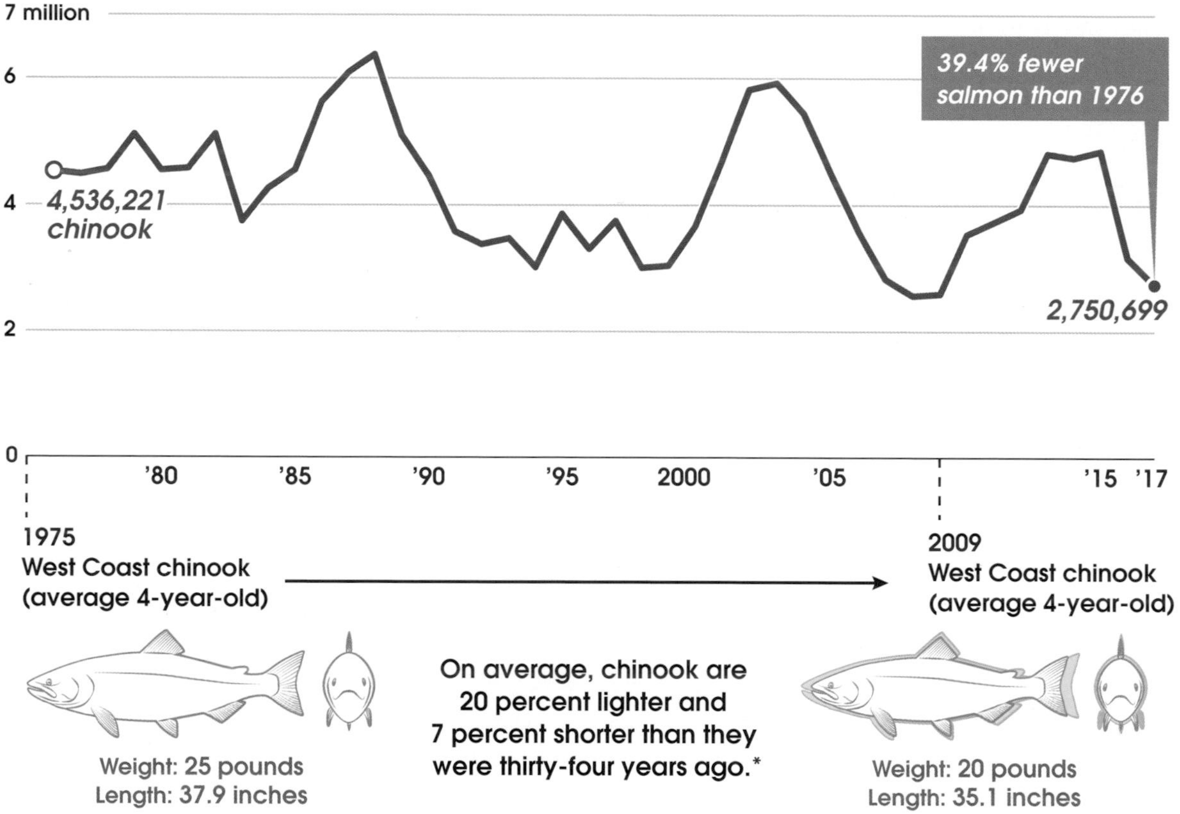

7 million
6
4
2
0

4,536,221 chinook

39.4% fewer salmon than 1976

2,750,699

'80 '85 '90 '95 2000 '05 '15 '17

1975
West Coast chinook
(average 4-year-old)

2009
West Coast chinook
(average 4-year-old)

Weight: 25 pounds
Length: 37.9 inches

On average, chinook are 20 percent lighter and 7 percent shorter than they were thirty-four years ago.*

Weight: 20 pounds
Length: 35.1 inches

*Weight and length measured for 4-year-old ocean chinook from multiple salmon runs from Alaska to California.

Sources: Jan Ohlberger et al., "Demographic Changes in Chinook Salmon Across the Northeast Pacific Ocean," Fish and Fisheries; Center for Whale Research; Pacific Fishery Management Council (2018); NOAA Technical Memorandum NMFS-NWFSC-123 (July 2013); Pacific Salmon Commission (2018)

EMILY M. ENG / *THE SEATTLE TIMES*

hatchery fish spend fewer years at sea. Hatchery fish also return to a building, not to the land. Unless humans hand-haul spawned hatchery carcasses into the woods, all the biomass that once fed the land and water is lost. Even in the good years when big runs come back, the food web is too depleted to support it. Scientists have determined that the number of fish returning to the rivers of the Pacific

Northwest is now so reduced that just *6 to 7 percent* of the marine-derived nutrients—nitrogen and phosphorous—once delivered by returning salmon is currently reaching Northwest streams. What fish that do still return are allocated in fisheries regulations to only fishermen. There is no quota to feed the rivers, the streams, the forests, the soils, the bears, the birds, the insects—or the orcas.

But perhaps the most damaging effect of hatcheries is the mind-set they engender: that habitat doesn't matter because we can just make more salmon. More insidiously, that salmon are a commodity to be manufactured, rather than a natural gift specific to a myriad of particular places that must be stewarded for the benefit of the fish.

Reliance on hatcheries has substituted for solutions to habitat damage, because hatchery fish prop up the fishing seasons they are geared to stock—albeit at pitifully low numbers. In some streams, hatchery fish are all the salmon that are left. In many rivers, including the Columbia, hatchery fish make up more than 80 percent of adult fish returns. Tribes and the southern resident orcas alike depend today on hatchery fish.

HOW MANY FISH DO ORCA NEED?

Chinook are the foundation of resident killer whale diets. Based on orcas' individual caloric needs, researchers have calculated the minimum number of chinook needed to sustain the current populations of resident killer whales.

The **75 southern residents** would need at least ***317,000*** chinook per year to survive with a diet of only chinook. A recovered population would need at least 554,000.*

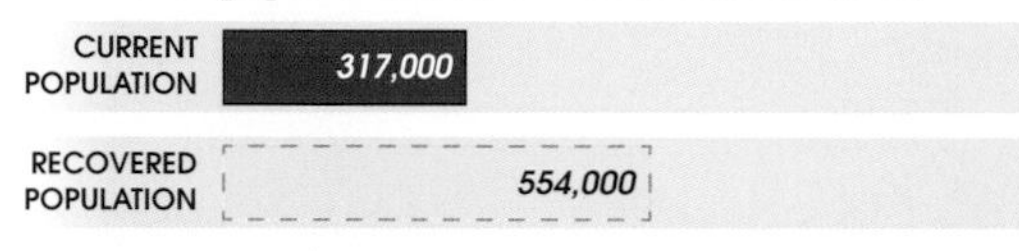

The **307 northern residents** would need at least ***1,150,000*** chinook per year to survive with a diet of only chinook.

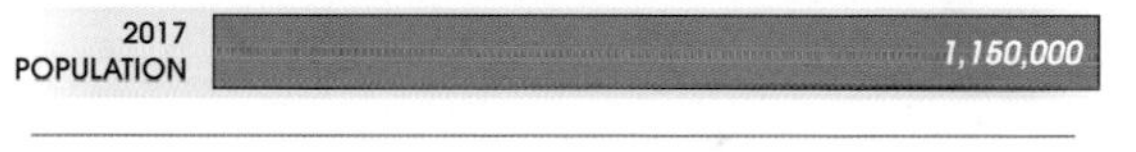

Based on preliminary data in 2018, ***871,292*** chinook were caught in all commercial, tribal, and sport fisheries from Alaska to Oregon.

2018 CATCH: *258,072* from Alaska; *360,354* from Canada; *252,866* from WA/OR; *871,292 total*

*Based on Rob Williams et al., projection of a 75 percent increase in diet need.

Sources: Fisheries and Oceans Canada; Pacific Salmon Commission; Rob Williams et al., "Competing Conservation Objectives for Predators and Prey: Estimating Killer Whale Prey Requirements for Chinook Salmon," Plos One

EMILY M. ENG / *THE SEATTLE TIMES*

Fewer Fish, Shorter Seasons

Exhausted, beat up, and underfed after 150 years of habitat destruction and the loss of the nutrition that salmon provide, the lands and waters of the Pacific Northwest no longer produce salmon like before. And it all happened so quickly.

Ron Warren, the fish program manager, remembers vividly the first salmon he ever caught: a 36-pound chinook, hooked at the mouth of the Columbia in 1968. He was just a little kid fishing with his grandfather, his dad, and his uncle. "We were all out in a boat together, and I still have that picture of me holding up that very first chinook salmon. I was pretty proud. It was an amazing day and memory I will cherish in my older years. I tell my kids about it, how much it meant to be able to fish with my grandpa and my dad. It's why I went to work at the Department of Fish and Wildlife."

It's been downhill ever since, with fewer fish and shorter seasons, and even the fish themselves are smaller. The giants that used to lumber up and down the Columbia and cruise the North Pacific

LEFT: Crystal Chulik inspects a chinook during scientific sampling at the Bonneville Dam. Big fish like this one are what southern resident killer whales need for a square meal but most chinook are much smaller than they used to be. The causes are many, including selective fishing for large chinook by people and predators alike. *(Steve Ringman/*The Seattle Times*)*

OPPOSITE: On NOAA's research vessel, the *Bell M. Shimada*, scientists sort a catch from their survey off Washington's coast. The surveys, conducted since 1998, help scientists understand factors influencing the ocean survival of juvenile salmon. *(Steve Ringman/*The Seattle Times*)*

from California to western Alaska have shrunk, Jan Ohlberger of the University of Washington's School of Aquatic and Fishery Sciences and other authors found in a 2018 paper published in the journal *Fish and Fisheries*.

This salmon shrinkage had been reported by researchers for years, beginning with William Ricker of the Department of Fisheries and Oceans Canada. He reported in 1980 that the average size of chinook salmon caught in the eastern Pacific Ocean had been declining since at least 1920 and continued to decline to average weights already half or less than half of that attained fifty years earlier. The slide toward smaller fish has continued in a widespread trend in both wild and hatchery chinook.

Coast-wide, the weight of four-year-old chinook on average has dropped by 20 percent from 1975 to 2005. June "hogs," as the legendary giant summer-run chinook of the Columbia were known, tipped the scales at 80 pounds as recently as the 1920s. Today they exist only in historic photos. A sampling of chinook caught in Washington from 1970 to the present by purse seine and troll gear indicates puny average weights, ranging from around 10 to 15 pounds. That's just a snack for a 6-ton orca.

Orcas utilize the rock walls, coves, and bays of the west side of San Juan Island to hunt salmon headed for the Fraser River. The noise and disturbance from ships and boats make it harder for orcas to catch already scarce prey. *(Steve Ringman/*The Seattle Times*)*

Perhaps not surprisingly, the orcas born in the era of smaller salmon have shrunk too. In aerial surveys done with drone photography, scientists have documented that younger orcas in both the northern and southern populations are shorter in length than older orcas born into these once-abundant seas. So hunger stunts even the ocean's top predator, John Durban and Holly Fearnbach found in a 2019 paper. Other survey work by the team shows that the southern resident orcas have poorer body condition overall compared to the northern residents and that their condition is deteriorating over time, resulting in thinner orcas.

The southern residents also have more competition for every fish than ever—but not so much from fishermen, who have been pushed to the side since the commercial and recreational fishing bonanzas of a generation ago. Now it is other animals, especially marine mammals on the rebound since the federal Marine Mammal Protection Act was adopted in 1972, that are mostly challenging the southern residents to a meal. On a bright winter day I went to the Nisqually River in southern Puget Sound to meet some of the competition.

Willie Frank III is a son of the late Nisqually elder Billy Frank Jr., who was arrested dozens of times in the 1960s and '70s by state game wardens as he fought to defend his people's treaty-protected fishing rights. Willie met me on the shore of the Nisqually River in southern Puget Sound to explain his frustration with seals and sea lions catching scarce and precious winter chum. We were standing side by side watching the Nisqually slide toward Puget Sound, whirling and sparkling, when suddenly a sleek brown head popped up.

The sea lion surfaced with a big chum salmon clamped in its jaws, shaking its head violently, sending chunks of fish flying, and diving underwater to get the pieces. Back up in minutes, the sea lion tipped its head back like a sword swallower and downed the rest of its meal.

Sea lions never used to come up this river, said Frank, a member of the Nisqually Tribal Council. Today, seals and sea lions travel more than 20 miles up the Nisqually after chum.

These are not just any fish. These chum are unique, among the latest winter salmon runs in the state. They are the prime fish that the southern resident orcas hunt when they come to central Puget Sound in winter. But this chum run has declined so much that tribal members barely get a fishing season anymore, said Frank. The exercise of treaty rights is at risk as salmon disappear.

Frank sees a parallel between the tribal elders and the southern resident orcas, both struggling to find enough fish. This was the fresh salmon that was on tribal tables from Thanksgiving until Christmas and that fed the southern residents as the Salish Sea summer chinook runs wound down. "To see the little ones out there, and their moms, it breaks your heart," Frank said of the orcas.

The population boom in marine mammals—other than southern resident orcas—is complicating the recovery picture, as everything from seals to sea lions and Alaskan and northern resident orcas beat the southern residents and fishermen to the catch.

A paper published in 2017 was a shock to many, when Brandon Chasco and other researchers showed that the resurgent population of marine mammals, thanks to the ban on hunting enacted in the federal Marine Mammal Protection Act in 1972, may have had unintended consequences. Along the entire West Coast from 1975 to 2015, the catch of chinook salmon by marine mammals increased by 50 percent and the total catch by all anglers, commercial and recreational, went down 41 percent. Whether to cull marine mammals is under region-wide debate.

Declining Salmon Diversity

The discussion of hunger and orcas has tended to be oversimplified. It isn't the case that all the southern resident orcas are starving. What *is* true is that too often the southern residents are food limited, meaning they can't reliably always get enough to eat. That is due not only to declines in overall abundance of many chinook runs but also to the loss of diversity of when and where those fish are found, a crucial factor often overlooked—and much harder to fix.

Of 396 populations of chinook that used to be available to southern residents all over the Northwest and California, 159 today are gone. That leaves gaps in the calendar year in which the southern resident orcas' preferred prey is no longer available. Chum also are depleted, with 23 of 112 populations no longer there. With so much diversity lost, not only in geographic distribution of the orcas' food but in the timing of when it is available, recovering the southern resident population isn't just a matter of pumping up existing stocks, said Mike Ford, director of the conservation biology division at NOAA's Northwest Fisheries Science Center in Seattle.

For instance, in the Columbia over the past twenty years, fall chinook runs have mostly been doing better than in the previous sixty or seventy years. Yet the orcas continue to decline. That's because the southern residents need salmon year-round, throughout their home range. And spring chinook, a prize meal arriving fat from the sea and vitally important in a hungry time of year throughout the Northwest, are among the most depleted salmon populations, including in the Columbia Basin.

There's no rescue underway that is right-sized to the southern resident orcas' food problem, according to Andrew Trites, director of the Marine Mammal Research Unit at the Institute for Oceans and Fisheries at the University of British Columbia in Vancouver. Fixing just one place or piece of the problem will never save the southern residents. "They live in a very large house, and we need to look at every room," Trites said.

Just pumping out more hatchery fish of the types already dumped into rivers by the hundreds of millions doesn't touch the problem of the populations of salmon that are no longer there; in some cases, reintroduction will be necessary, said Mike Ford. The good news is that reintroduction can work. Snake River fall chinook are one of the few success stories in the Columbia Basin. They surged back to relative abundance from an experimental population planted after the run dwindled to nothing. Today those fish are one of the strongest amid chinook runs in the Columbia-Snake River system, and the southern residents do eat them.

But what about the crucial runs in the Fraser River presently in steep decline, the California Central Valley fish runs fallen to near extinction, and hundreds of other populations already lost? This beautiful diversity of salmon-run timing and life history—when and where salmon emerge and where they come home to—is a crucial survival strategy for the salmon and the orcas alike. It means that in the event of a bad year in a home tributary, other age classes of the same type of fish are safely out at sea. Variability of ocean conditions also can be balanced by fish with different migration paths and timing. Spatial diversity matters, too, with fish coming back to streams all over a broad geography, from Alaska to California, to lower river reaches and high mountain redoubts, again spreading the risk and giving the orcas and the salmon the edge of multiple opportunities. This diversity is what the orcas coevolved with. They are adaptable; they have learned to go where the fish are as seasons and conditions change. But they have to be able to reliably find fish *somewhere*.

Brad Hanson, a research wildlife biologist with NOAA's Northwest Fisheries Science Center, said people forget about how much the baseline for salmon and orcas has shifted—and how fast. "If you look at all the areas the whales take fish out of, it's a huge swath of North America, all the way to B.C. These animals evolved to depend on all these different stocks," Hanson said. Today, scientists are concerned about what they call seasonal serial failures: when, from one season to the next, in one river after another, there is not enough food regularly available for the southern residents. "If California is bad and the Columbia is bad and the Fraser is bad, that takes out six or eight months of the year," Hanson said. "You are not going to make it. You are potentially losing calves or individuals, and that is what we are seeing."

Orca grandmother J17 died in 2019 after losing so much fat that the curve of her neck showed, a deadly condition called peanut head. Losing J17, age forty-two, also put other whales in her family at risk because they relied on her to find and share food. *(Dave Ellifrit/Center for Whale Research; taken under NMFS Permit 21238 and DFO SARA Permit 388)*

Scat Tracking

Hunger hurts and even kills: the southern residents have a high rate of failed pregnancies. In 69 percent of pregnancies tracked from 2008 to 2014, no live calf was produced, according to a 2017 study led by Sam Wasser, director of the Center for Conservation Biology at the University of Washington. Wasser documented a connection between failed pregnancies and stress hormones in the southern resident orcas' scat and periods of low salmon abundance in the Columbia and Fraser Rivers.

The orcas' reliance on chinook has been documented mostly through bits of prey scooped up in a net by researchers following the orcas in inland waters and also from scat samples. I've been on several of those surveys as a reporter and have by now experienced a surprising array of intimate orca excreta. I've seen orca snot, orca scat, shreds of salmon and glittering scales from an orca kill, and bits of slippery black orca skin, all gathered in researchers' nets. I've poured orca scat into vials and gotten just as excited as the researchers at a nice big sample, glistening with globules of fat: a sure sign of a healthy orca.

Some researchers follow the orcas to find the scat. Others also bring trained dogs aboard to help them, which, with the exquisite sensitivity of a dog's nose trained to the task, allows researchers to hunt down the tiniest bits of scat even a long distance away from the orcas. I learned that watching orca scientist Deborah Giles doing surveys for Wasser's study with Dio, a dog from the lab's Conservation Canines program trained to find orca scat.

On a radiant July day in 2018 we set off from the dock on Giles's small open boat, *Moja*, from the west side of San Juan Island, a favorite southern resident orca summer foraging ground. Recreational whale-watch captains already out on the water had confirmed there were orcas out at the south end of the island. Giles opened the throttle and we headed out to prospect for scat. Once she found the orcas, she followed their fluke prints—large glassy patches on the surface, created by the movement of the orcas' tails as they swim along—guided also by the acute nose of Dio, a blue-heeler mix.

Handled by trainer Collette Yee, Dio was one of the dogs, all of them rescues, in Wasser's Conservation Canines program, crack environmental

TOP: Dio and his handler Collette Yee of Conservation Canines search for orca scat in the waters off San Juan Island. *(Steve Ringman/*The Seattle Times*)*

BOTTOM: Scientist Deborah Giles examines a sample of orca scat. Researchers have documented a link between stress hormones in orca scat, low salmon availability, and failed pregnancies. *(Steve Ringman/*The Seattle Times*)*

detectives trained to track everything from invasive plants to polychlorinated biphenyls (PCBs). Dio was a muscular animal, taut as a wire and focused on his task. Yee watched the dog so she could give hand signals to Giles, who was steering the boat. Giles's job was to be the dog's legs, moving the boat to where his nose pointed.

Before long, Dio located a particle of scat. Giles scooped it with her net, then set it spinning in a vial in the shipboard centrifuge, for analysis back at the lab. I held the whizzing centrifuge with my foot to keep it from shimmying across the deck.

This sample would tell researchers everything from what the orcas were eating to the orcas' condition and, using DNA analysis, the species and origin of the fish. "Within four days we see the impact if they are not getting enough nutrition," Giles said. "Any animal goes through feast and famine; that is normal. But their periods between feast and famine are bigger."

She was sad at the sight of this scat—small, tight, and dry. Where was the size, the fat, opulent traces of a big meal? She and other researchers were also hoping for more samples. But the orcas were hardly coming around, a pattern the scientists figured was linked to the downturn in Fraser River fish the orcas pursue while in their summer habitat, which includes the San Juan Islands. It used to be that the orcas returned regularly to the San Juans in May and were around nearly every day. Lately, the southern residents were more typically arriving much later and were split up and spread out, with only a few of the families together in any one location. They were socializing and resting less, traveling more: looking for food.

A Fast-Changing Environment

In this way, lack of food affects not only the orcas' health but their social lives and family cohesion—just as it does human families. Where are the elders out catching fish with the next generation, as Ron Warren's elders did with him? Increasingly, these are memories—in picture frames for people and in the minds of the southern resident orca matriarchs, remembering an abundance that used to be.

"The environment has changed so quickly, over two generations, this population's ability to keep up with these changes is almost an impossible task," said Sheila Thornton, lead orca scientist for Canada's Department of Fisheries and Oceans. "How do you survive in this environment we have created for them? They are not fasting adapted; they have to eat every day."

Thornton said she thinks what might be most important to the orcas' recovery is rebuilding the stocks that are no longer there—hungry gaps in the seasonal year. In their quest for food, the orcas tell us what's wrong with the bigger picture, in a region of abundance no longer able to produce enough food for the ocean's top predator.

The freshwater streams and rivers that produce salmon hundreds of miles inland are just as important to southern resident orca survival as the marine environment they also depend on. "It can be hard for people to understand. They see a stream that feeds orcas that is 60 kilometers inland, but everything is connected," Thornton said. "When an apex predator is showing signs of decline, people have to look at everything that supports them. This is actually about where *we* live."

The primal task of hunting salmon is also getting harder for the orcas. "It's not just the abundance of chinook salmon but the urbanization. There is so much motorized vessel traffic and fewer fish, they are no doubt having a harder time detecting them acoustically," said John Ford, the Canadian orca expert. "You put a hydrophone [underwater microphone] out there, you listen to that din out there; it is rare there is any kind of quiet period. It makes it harder for these animals when they are widely dispersed and foraging to coordinate their activity, and there are ultrasonic aspects that may interfere with echolocation and make it more difficult to detect salmon in their midst."

Because of our racket, the southern residents are forced to hunt not only in the dark but in a fog of noise.

CHAPTER FOUR

The Roar Below

In an ancient drama of predator and prey, orcas that frequent the Salish Sea prowl the waves and dive glacially carved fjords and bays, undisputed masters at hunting the salmon they coevolved with.

Like fishermen everywhere, the J, K, and L pods of southern resident orcas have deeply set patterns of how, when, and where they hunt, depending on seasonal salmon migrations, tides, and the underwater

The southern residents live in increasingly noisier and more polluted water in their urban habitat, including the waters of central Puget Sound. These orcas swam between the Port of Tacoma and the Superfund site at the former Asarco smelter at Ruston in November 2018. *(Steve Ringman/*The Seattle Times*; taken under NOAA Permit 21348)*

landforms they use to capture a wily target. But in some of their ancestral hunting grounds, the southern residents are losing out in a clash of two great maritime cultures: orca and human.

Right where the southern residents have learned, over thousands of years, to use the rock canyon along the west side of San Juan Island like a fish funnel to nail chinook returning to the Fraser River, humans have in just the last century created an echo chamber of industrial noise. Haro Strait is a long stretch of water between British Columbia and San Juan Island that serves as a main drag for shipping traffic to and from Vancouver. It booms with ships, ferries, recreational boats, and whale-watching tours.

More than 300,000 British Columbia and Washington State ferries traveled the Salish Sea in 2018, while 6,330 cargo, container, and passenger vessels and 1,134 oil tankers and barge tows also entered Washington waters. Much of that traffic is headed to the Port of Vancouver, the biggest port by cargo tonnage on the West Coast. Ships are present in Haro Strait in every season, day and night. Much of their noise is in the same sonic sweet spot that orcas use to hunt and communicate.

That makes the orcas' prey, already scarce, even more difficult to catch. Scientists have learned that orcas forage less in the presence of vessels, switching from foraging to traveling if they are closely approached by boats. The orcas also have to call louder to be heard over anthropogenic noise of every sort, from fish finders to boats to ships and construction, costing them precious energy.

"We are asking them to do all these things in a way that is totally a game changer now," said Marla Holt, research wildlife biologist at NOAA's Northwest Fisheries Science Center in Seattle. "They are trying to

THE SOUTHERN RESIDENTS' NOISY HOME

The endangered southern resident orcas that visit Puget Sound confront the noisiest waters in their critical habitat, including the west side of San Juan Island, the Fraser River delta, and the Strait of Juan de Fuca. Noise is caused by vessel traffic, especially commercial shipping. Their habitat in all of the Salish Sea has underwater noise levels that do not comply with the noise-pollution limits recommended by the European Union.

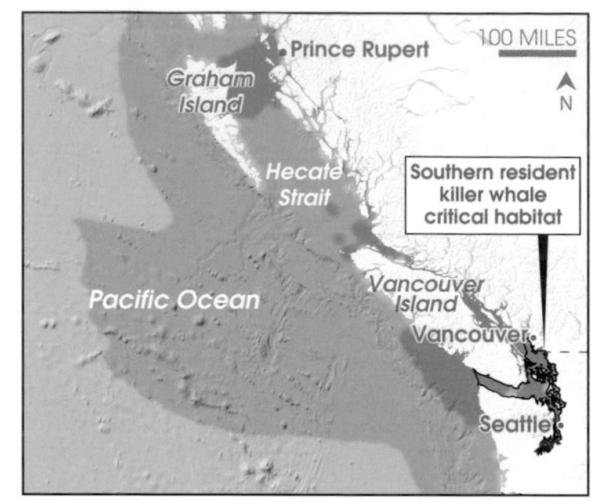

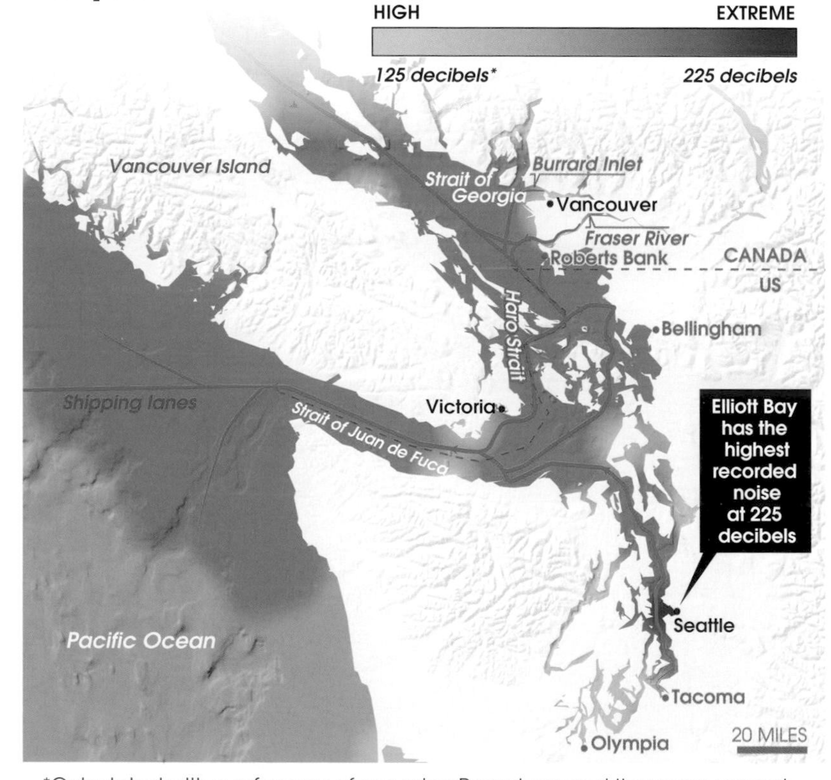

*Calculated with a reference of one microPascal squared times one second

Sources:* *Christine Erbe et al., "Mapping Cumulative Noise from Shipping to Inform Marine Spatial Planning,"* The Journal of the Acoustic Society of America; *Washington State Department of Ecology

TOP: Scientist Rob Williams, left, of the nonprofit Oceans Initiative, and Joe Gaydos, senior scientist of the SeaDoc Society, listen to the racket of industrial shipping underwater using a hydrophone dropped into Haro Strait. *(Steve Ringman/*The Seattle Times*)*

BOTTOM: Retired physicist Val Veirs prepares to put a listening device in the water offshore of his home on San Juan Island to record the noise of ships, ferries, and boats. *(Steve Ringman/*The Seattle Times*)*

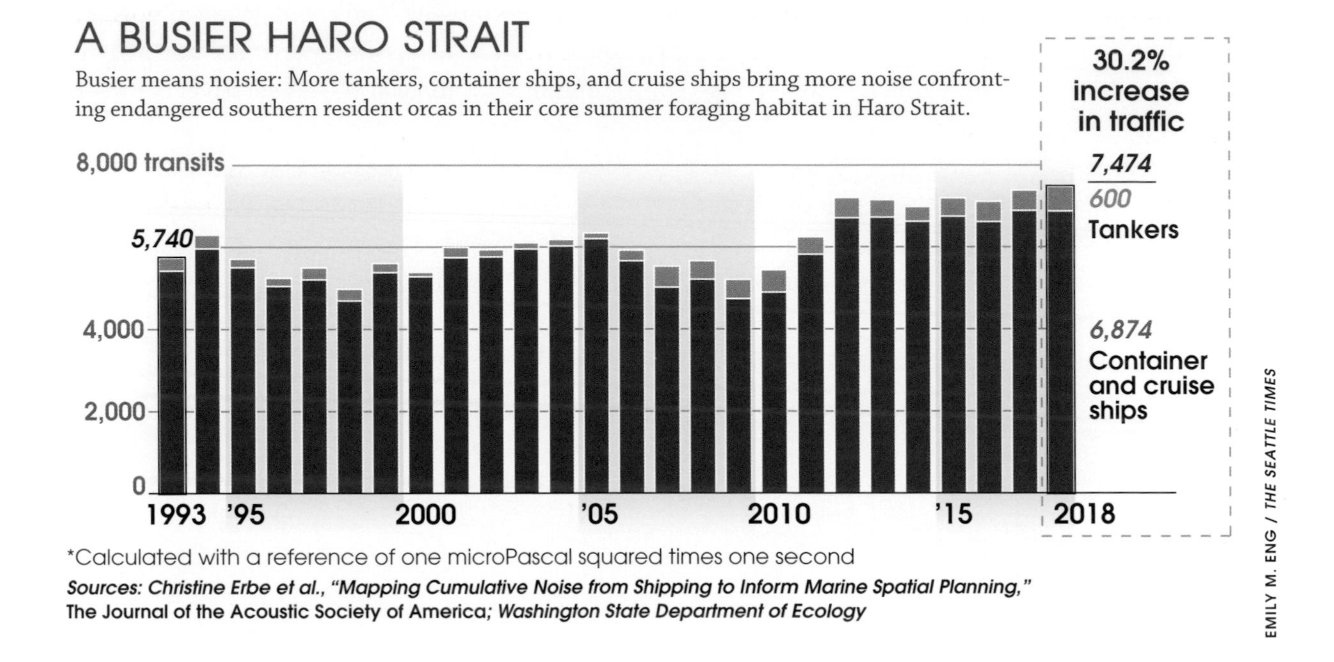

EMILY M. ENG / *THE SEATTLE TIMES*

find these rarer and rarer fish. It would be like going grocery shopping and I am going to turn out the lights and you only have twenty minutes and I am going to take half the food away—good luck with that."

Listening In on the Orcas' World

To get a sense of what orcas cope with in their primary summer foraging ground, I joined up again with Joe Gaydos of the SeaDoc Society and Rob Williams of Oceans Initiative on the *Molly B* to experience a typical day on the west side of San Juan Island—with a hydrophone (an underwater microphone) tossed in the water to hear underwater noise. We didn't have long to wait before the first ship of the day came into view.

With a rhythmic thudding, the container ship heading from the Port of Vancouver to the Strait of Juan de Fuca and the Pacific powered into the north end of Haro Strait and down the west side of San Juan Island. Williams had the hydrophone playing the ship's noise over a speaker.

Gaydos had heard the ship on the hydrophone from 10 miles away, long before he saw it. "It looks like a small city," he said as the ship, the *Xin Los Angeles*—flagged in Hong Kong—barreled along at 22 knots. With a capacity for 9,600 containers stacked seven rows high, it is nearly twice as long as the Space Needle is tall, the longest container ship in the world when built in 2006. The ship obliterated the view of Lime Kiln Lighthouse and left a seething wake.

It was an ordinary moment in Haro Strait, but an extraordinary disturbance for animals sharing the waters and soundscape. Loafing harbor porpoises chuffing at the surface scrammed when the ship showed up.

As the Northwest grows and its economy booms, with the closest ports to Asia, not a day—and scarcely an hour—passes without heavy industrial shipping traffic, recreational boaters, fishing vessels, and ferries transiting critical habitat of the southern resident orcas, including their prime foraging grounds in the summer.

During our five-hour survey on the *Molly B*, six ships traveled Haro Strait, flagged in four nations besides the container ship *Xin Los Angeles*: one vehicle carrier, two cargo carriers, one bulk carrier, and one Canadian military training vessel, the steel-hulled *Renard*—dubbed Orca Class—with a rooster tail of water shooting behind it. A blazing-fast crab boat was so far away it took binoculars to see but could be heard for miles.

The Salish Sea is a major maritime hub. The Northwest Seaport Alliance, comprised of the ports of Seattle and Tacoma together, is the fourth-largest container gateway in North America, with $73 billion in international trade in 2017, according to the alliance. In Washington State alone, the maritime sector employed nearly 70,000 workers and generated more than $21 billion in revenue in 2015, according to the state Department of Commerce.

The Port of Vancouver sustains trade with more than 170 economies around the world and supports more than 96,000 jobs in British Columbia alone. In 2018, 147 million tons of cargo moved through the port, valued at $200 billion. The Port of Vancouver saw record-breaking traffic in 2017 and again in 2018. Current demand forecasts in a study for the Port of Vancouver anticipate container trade to double in the next ten to fifteen years and nearly triple by 2030.

Waters already loud could get even noisier. The Port of Vancouver wants to build another major container terminal at the Fraser River delta—right where orcas hunt. That proposal is under environmental review by the government of Canada, which also is poised to increase oil tanker traffic sevenfold in Burrard Inlet to serve a controversial expansion of the TransMountain Pipeline, boosting traffic from about 60 to more than 400 oil tanker trips per year. All of those oil tankers would get to Canada through the southern residents' primary summer foraging grounds.

Impacts of Noise

Underwater noise can affect the southern residents by changing their behavior, displacing them from their range, masking their communication, decreasing their foraging efficiency, creating stress, and even damaging their hearing.

Overall, southern residents potentially lose up to five and a half hours of foraging time each day from May through September because of vessel noise, researchers have found, with approximately two-thirds of those effects due to large commercial vessels and one-third due to whale-watching boats.

Oceanographer Scott Veirs of Seattle and his father, Val Veirs, a retired physicist on San Juan Island, found that large ships have the biggest influence on the orcas' ability to hear, both because ship traffic occurs year-round and because ship sound frequencies overlap with the orcas' hearing range.

Scott Veirs maintains the OrcaSound network of hydrophones in the Salish Sea, including right off the beach on the west side of San Juan Island. Sounds stream in: the breathing of harbor seals, groans of fish, even the drone of airplanes overhead,

LEFT: Whale watchers from British Columbia observe transient orcas. In an industry that draws about five hundred thousand people a year, southern resident orcas now comprise only 10 to 15 percent of sightings. Most whales seen are humpbacks, gray whales, and transient killer whales. *(Steve Ringman/*The Seattle Times*)*

OPPOSITE: Speed is a top factor in the amount of noise orcas receive from vessel traffic. This whale-watching boat is headed back to port after an evening cruise along the west side of San Juan Island. *(Steve Ringman/*The Seattle Times*)*

heard through the water. But mostly what they record is the sound of ships.

In a paper published in 2016, Scott Veirs and his coauthors found that ships are raising noise levels right in the same frequencies of orca hearing. "The most complicated thing they [orcas] do is hearing the echo off the swim bladder on a chinook that is fifty to a hundred meters away in dark, cloudy water, where they can't see more than a whale's body length," said Scott. "They are masters of sounds, and the entire heads are miraculous mechanisms for both generating sound and receiving it. But they need to be able to receive a very faint sound, the echoes off a little organ full of air inside a salmon."

In addition to finding prey, southern residents need to hear one another. The southern residents have their own unique dialect, and each pod has favorite calls. The southern residents are known for their chattiness, carrying on an unending conversation with one another. For researchers who travel with the southern residents, it's not uncommon to hear their calls bubble up to the surface.

Although big ships have been singled out as the noisemakers that matter most, whale-watching and other small-boat traffic can have an effect, depending most of all on their speed of travel. Speed is the single most important predictor of sound that will be received by the orcas, Holt and her coauthors found in a 2015 paper.

By 2020, the whale-watch industry had grown to twenty-eight companies in nineteen ports in Washington and British Columbia, drawing more than half a million people every year from all over the world who are thrilled at the possibility of seeing whales. Business grew even as the first federal and state restrictions intended to reduce impact on the southern resident orcas were implemented.

UNDERWATER SOUNDS

As orcas chase prey into deeper, darker, and more turbid waters, their ability to see declines. So orcas use sound as their primary sensory system for communication, navigation, and finding prey.

TYPES OF SOUNDS

Killer whales make three different types of sounds: whistles, calls, and clicks. Whistles and calls are used for communication, while echolocation clicks help with navigation.

Whistles are continuous single tones at high frequencies used for close-range communication. Calls are made in lower frequencies and can travel up to 9 miles. Pulse calls create rapid streams of sound and are the most common vocalization. These calls are used for finding and staying in contact with one another and coordinating movement.

Calls that always sound the same are called discrete or stereotyped calls. Pods that share a number of discrete call types form an acoustic clan with its own unique vocal tradition. Even within a clan, different groups can have their own unique way of making certain calls and forming their own dialects. These dialects are distinct enough that no two are the same.

Sounds are so important to whales that within days of birth calves begin to vocalize. Around two months old, calves can send pulse calls similar to the adults. They selectively learn calls from their mother as they mature.

WHISTLE
Frequency: 2,000 to 50,000 Hz
Duration: 60 to 18,000 milliseconds

PULSE CALL
Frequency: 500 to 30,000 Hz
Duration: 600 to 2,000 milliseconds

ECHOLOCATION CLICK
Frequency: 10,000 to 100,000 Hz
Duration: 0.1 to 25 milliseconds

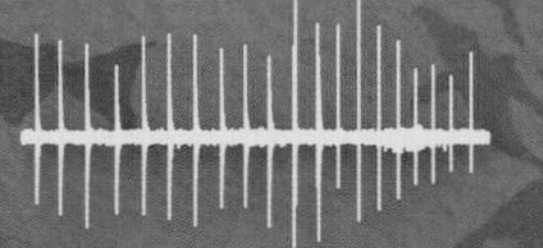

Sources: Marla Holt, NOAA Fisheries Northwest Fisheries Science Center; Scott Veirs et al., "Ship Noise Extends to Frequencies Used for Echolocation by Endangered Killer Whales," PeerJ

SOUND VS. NOISE

A sound is an acoustic signal that is important to the listener. Noises are unwanted sounds that interfere with the reception and transmission of sounds. Noise can affect different qualities of sound, like the frequency (rate of vibrations measured in hertz), loudness (amplitude in decibels), or duration. A sound wave can travel almost a mile per second underwater—more than four times faster than in air, with low frequencies traveling farther than high frequencies.

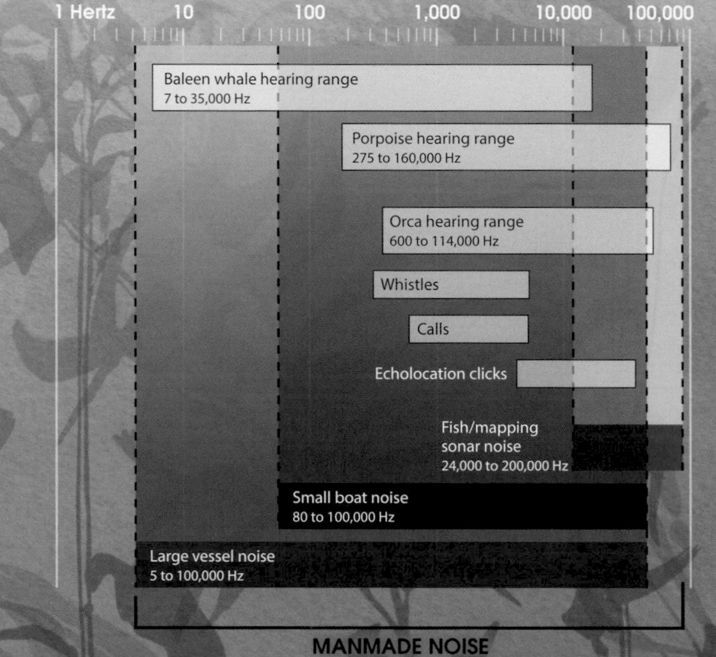

VESSEL NOISE

Between 80 to 85 percent of vessel noise is generated by ship propellers. The rest is created by propulsion machinery, including the engine, and by water flowing over the hull. Large vessels create lower frequency noise that can travel hundreds of miles underwater. Noise also increases with speed and proximity.

ACOUSTIC MASKING

Since the 1950s, underwater ambient noise has doubled each decade (three decibels every ten years). This forces orcas to increase the loudness (one decibel for each decibel of noise) or length of their calls. Being louder comes at a cost of increased energy required for sound production and increased stress levels. A noisier environment also decreases the distance at which orcas can detect prey, causing whales to work harder to find food.

Many salmon runs also already are depressed. In this way the threats to orca survival combine and interact to cause a greater overall peril: Noisier water makes it harder to find increasingly rare fish, which results in bigger exposure to pollutants for whales that don't get enough to eat. Hungry whales burn the fats in their bodies where pollutants are stored. That exposes them to toxics that harm reproductive capacity and reduce their ability to fight disease.

EMILY M. ENG / *THE SEATTLE TIMES*

Autumn still means orcas in Puget Sound country, when the southern residents come back home. These onlookers at Point Robinson, Maury Island, were thrilled in November 2018 as the J, K, and L pods frolicked right offshore, just south of downtown Seattle. *(Steve Ringman/*The Seattle Times*; taken under NOAA Permit 21348)*

The industry used to depend on the southern residents for its bread and butter. Today, a surge in humpback whales and marine mammal–eating transient orcas keeps the customers coming, said Jeff Friedman, US president of the Pacific Whale Watch Association. Currently, the southern residents make up only 10 to 15 percent of the whale-watch business because they are here less. One likely reason is the decline in their primary food: the chinook returning to the Fraser River.

When Threats Combine

"If there is plenty of food, maybe vessel noise doesn't matter that much," Holt said. "And that is where all these risk factors intersect." Noise isn't the sole issue; physical disturbance also matters. "It is a combination of things, keeping track of all these other obstacles in my environment; especially if I have to move very quickly . . . those physical challenges are going to be an extra challenge."

Some noise is overwhelmingly disruptive. Ken Balcomb, founding director of the nonprofit Center for Whale Research on San Juan Island, Washington, filmed an infamous incident in 2003 when the navy vessel USS *Shoup* let loose piercing bursts of sonar that sent J pod fleeing nearly up onto the beach, apparently trying to escape the noise. The incident was a black eye for the navy. Its testing program, including conducting sonar exercises and detonating explosives, remains controversial.

Balcomb is convinced it is the lack of salmon driving the southern resident orcas to decline, not ordinary boat and vessel traffic. He remembers when there were more fish in these island waters—and more fishermen chasing them. "It was like a village out there, full of lights, seal bombs going

NO SMOKING

IVORY ARROW
NASSAU
CHINA SHIPPING LINE

CLOCKWISE FROM TOP LEFT: The Haro Strait is home to an industrial shipping channel right in the middle of prime summer foraging territory for the southern resident orcas. *(Steve Ringman/The Seattle Times)*; Car carrier *Ivory Arrow* is one of many ships plying the Haro Strait, the main route to the major shipping port of Vancouver, British Columbia. *(Steve Ringman/The Seattle Times)*; Sharing their ancestral waters with ships, ferries, fishing vessels, naval operations, recreational boaters, and whale-watch tours forces the southern residents to compete with human noise and disturbance in one of their most important feeding areas. *(Steve Ringman/The Seattle Times)*; Ferries are yet another source of noise and disturbance for southern resident killer whales in both Washington and Canadian waters. *(Candice Emmons/NOAA Fisheries; taken under Permit 16163)*

off, boats everywhere," Balcomb said. He has drone footage of orcas traveling the waters near his house on the west side of the island without a flinch as a clueless boater blasts close by. "They are acclimated," he said.

Another explanation, according to Rob Williams of Oceans Initiative, is that it depends not only on the noise but on what the orcas are doing. The orcas are bothered by boats more when they are feeding. He and other coauthors reported in a 2009 paper that the southern resident orcas they monitored were more likely to stop foraging when they encountered a boat but continued traveling in the presence of boats, taking little notice.

In a 2013 paper, Williams and other coauthors also found that the areas where orcas can effectively communicate were greatly reduced by noise, with their world shrunk in the loudest places to just a few usable spots. Reducing vessel noise to increase orca hunting efficiency is one thing people can do right away to buy time for the southern residents, while also working hard to build up chinook runs. "Noise is a problem," said Williams, "because lack of chinook is a problem.

"We are trying to save the final seventy-five," he said of the southern residents, at their lowest population when he made that remark in 2018 since the capture era in the 1970s. "And we need every tool in the box. This population of whales is critically small. It has nothing left to give."

Yet not long after Williams's statement, the population of southern residents dropped still lower. That has everything to do with the condition of the habitat in which the southern resident orcas and chinook are struggling to survive.

CHAPTER FIVE

Hostile Waters

If there is a hell for salmon, this is it.

I looked dubiously around the open skiff where I would be spending the day with Doug Killam, senior environmental scientist for the California Department of Fish and Wildlife on a spawning survey of the Sacramento River. It was a June morning in the summer of 2019, and by ten in the morning, the sun had already raised the temperature above 100°F. I noticed a

Shasta Dam, finished in 1945, was built without fish passage. Water withdrawals, drought, and warming water—all worsened by climate change—have pushed Sacramento River winter chinook, an important food source for southern resident orcas, to the brink of extinction. *(Steve Ringman/*The Seattle Times*)*

half-gallon jug of frozen water Killam had brought aboard was already melting.

As we got underway from a municipal boat dock in Redding, we saw many more golf balls in the water than salmon, whacked there by enthusiasts at Aqua Golf, a driving range along the river. Below the surface, the gravel that salmon need to make their spawning nests had been mined decades ago to build Shasta Dam, 602 feet tall and with no fish passage. The dam had cut off access to all the cold mountain waters where these fish used to spawn. The hillsides above the river were blackened by wildfire. Houses, instead of forests, stood along the banks. Cars roared by on Interstate 5.

Yet the matriarchs of the orcas that frequent Puget Sound still remember the big winter chinook that used to thrive here. The fat, juicy fish are precious winter food for the orcas at the southernmost end of their foraging range. These orcas are called southern residents for a reason. They cruise south all the way to California to feed on Central Valley salmon runs. L pod was off Monterey Bay in early 2019, with the oldest orca among all the southern residents, L25, leading the way. She brought her whole family here because her mother had before her and her mother before that. But was L pod chasing fish in California—or only L25's memory of them? California Central Valley salmon, and particularly winter-run chinook, unique in the world, have become so scarce that it is hard to know if the orcas got any nourishment.

Suddenly Killam said, "She's alive," and turned off the motor. We dragged the boat onshore and looked into the shallows, where a big female chinook held steady in the water, near a light patch of gravel: her redd. She was barely moving.

That summer was a good one for the winter run, relatively speaking: the best in about a decade, after rains eased a long and severe drought. But the return of some one thousand fish was still just a fraction of the abundance of one hundred thousand and more winter-run chinook that used to help sustain the southern residents. "She's just hours from death; she's done her job," Killam said of the catatonic fish.

I watched the big chinook fin quietly in the water. She could have been among the very fish that L pod was targeting when the southern residents were offshore that spring. A big male chinook shot past the boat, defending his chance at a mate.

The setting had a doomsday feel to it as we looked at this dying fish from a revenant run, in a river running on what's left of its flow through a landscape in many places torched by wildfire. These fish are among the salmon most endangered in the West by climate change, because of higher temperatures in both the air and the water, with even more extreme droughts predicted in the future. For Killam, the climate emergency is already here.

In the summer of 2018, he was dispatching survey crews in masks—to protect them against wildfire smoke—to count the carcasses of this critically endangered species. "The irony is not lost on me," Killam said. "It was apocalyptic and terrifying. We are at a tipping point. These fish can only hold on for so long."

At Livingston Stone National Fish Hatchery, a last-ditch effort to prevent extinction of winter chinook carries on. Here, winter-run chinook are spawned from fish caught in a trap below Keswick Dam, downstream of Shasta Dam. Raised entirely in captivity, these juvenile fish are released back

Winter chinook circle in tanks, part of the captive brood at the Livingston Stone National Fish Hatchery at the base of Shasta Dam. When it was built, the dam cut off all their natural spawning habitat. *(Steve Ringman/* The Seattle Times*)*

into the river in small batches each year in an effort to preserve the genetic stock of the fish, while trying not to overwhelm what's left of the naturally spawning run.

At the looming concrete monolith of Shasta Dam, guards opened a series of locked gates and allowed us to proceed to the fish hatchery. No tours here, no public allowed; this is an intensive-care ward for a critically endangered species. As we arrived, technicians from the hatchery came out to sort through the catch of winter-run chinook from the trap at Keswick. Their job was to find the fish to keep for the hatchery program and to truck the rest back to the river.

Beau Hopkins, a fish culturalist with the US Fish and Wildlife Service, which runs the facility, climbed a ladder to get into an open hatch atop the truck and started sorting. He took tissue samples for genetic analysis as he worked, crisply ripping a scale off each fish with forceps and punching a hole in its tail with a hole punch to retain a little circle of each particular life for genetic analysis. He held up a big keeper: meaty and spotted, gaping its jaws.

Winter-run chinook look like something from the Pleistocene, and they certainly are anything but pathetic. These are strong and mighty fish, and as if to prove it, one of them, as Hopkins had his head down in the truck, leapt out an open hatch. It hit the hot asphalt of the road, thrashing and leaping, as the hatchery workers yelled "Fish! Fish!" They scrambled to the rescue, hustling the fish back into the truck, appalled such a thing would happen in front of visitors.

Hopkins kept working, passing the keepers in a black rubber sleeve to waiting hatchery workers on the ground who carried the fish to holding tanks. There, the fish would continue to mature and ripen until the workers spawn them to rear a new generation.

The captive brood was held in tanks under a roof next to the hatchery, kept there as a hedge against

LEFT: Almonds are a thirsty crop. It takes more than a gallon of water to grow one nut. Irrigation for industrial-scale agriculture takes water from the Sacramento River that salmon need. (Steve Ringman/*The Seattle Times*)

OPPOSITE: Today most salmon runs in the Sacramento River are so diminished, many people have forgotten that California is a salmon state at all and the California coast remains essential orca habitat. *(Steve Ringman/*The Seattle Times*)*

extinction. Hopkins unzipped the curtains around one of the tanks and got inside with a flashlight to show me the fish. It was a creepy moment. These fish had been grown to adults in captivity. They were small and soft looking—nothing like the wild-reared salmon that had just flung itself out of the truck with one flex of its back.

Everyone I spoke to that day talked about their commitment to the work of saving this species and about an urgent sense of purpose. Meanwhile, looming high above and behind the whole operation was Shasta Dam. Seldom are cause and effect so visible: more than 600 feet high, it fenced these fish out of the cold-water habitat upstream they had always used for spawning. The dam was streaked with water leaking down its face—the water held back from these fish.

Programmed releases of cold water piped from the depths of the reservoir behind the dam are provided to enable winter chinook to spawn in the valley floor, where they never before did. But the amount of water available for fish keeps shrinking, as humans take more for cities, industry, and agriculture. Climate change is making the situation worse, with longer and more severe droughts and hotter air and water temperatures. Meanwhile, growers and the US Bureau of Reclamation want to raise the dam higher, to keep even more water for irrigation and other human uses.

The Central Valley's Thirsty Crops

Part of what threatens winter chinook isn't only human population growth and climate change. It is the transformation of agriculture in the Central

Valley not only in scale but in what is being grown. Farmers used to fallow their crops in drought years, which are part of the natural rhythm of the state's Mediterranean climate. But with the transition to high-value perennial crops, including almonds, irrigators demand water, no matter what.

On my way from San Francisco to meet Killam, I had driven alongside mile after mile of almond orchards. I pulled over for a closer look and felt the heat assault the air-conditioned tin-can rental car. I cracked open the door cautiously as if confronting a mugger. It was 99°F at eight o'clock in the evening. The irrigation water was hissing, the trees verdant, but outside the orchard rows, the ground was a dead zone. Within reach of the white plastic irrigation pipes, the trees were lush, green, and loaded with nuts. Almonds are California's biggest export crop. Some people call almonds the solid form of water: it takes more than a gallon of water to produce every nut.

With winter-run chinook headed to extinction, NOAA scientists have concluded the best thing to do for both the fish and the southern resident orcas that depend on them is to get winter chinook out of this mess in the Central Valley, where it's too hot and there's not enough water. Put them back up where they used to spawn, in the McCloud River, before they were cut off from the high country by Shasta Dam. Envisioned by NOAA is a trap-and-haul operation in which returning adult fish would be collected in the Sacramento River and hauled above the dam by truck to their original cold-water spawning grounds. Juveniles would be gathered above the dam and then trucked and released below it to the Sacramento River, to continue their journey to the sea.

Two studies of winter chinook by NOAA biologists found the relocation plan a preferred alternative to just continuing business as usual in the operation of Shasta Dam for Central Valley water projects. That, the agency's scientists found, would jeopardize survival of not only winter-run chinook salmon but also the southern resident orcas. Each has been designated by NOAA as a Species in the Spotlight—the ten listed species in the country that the agency regards at most urgent risk of extinction. But after decades of study, the $9 million relocation plan was iced by politics over Labor Day weekend 2019. "We were so close," said Jonathan Ambrose, reintroduction coordinator for the National Marine Fisheries Service in its Central Valley office, in Sacramento. What happens now is anyone's guess, Ambrose said.

Trap and haul is common in Washington State, used to move salmon past dams on multiple rivers, including the Cowlitz and the White. So engineering isn't really the issue; trap and haul is nothing crazy or new. The bigger challenge for winter chinook, according to Ambrose, is whether anyone in California even cares enough about salmon anymore to try to save them. "For a lot of us here, we feel like we are just screaming into the wind," Ambrose said. "People don't even know what we used to have. A lot of people don't even know we have salmon in our rivers. It's a huge disconnect.

"When you say 'salmon,' people think of the Pacific Northwest. One of your advantages in the Pacific Northwest is, people care. But here, the fish have been down so low for so long, it's hard for people to care about something they have never seen. We have driven winter chinook to near oblivion, and people don't even know about it. How sad is that? We are fighting to keep them, but essentially most of the runs are listed for protection. A lot of people have come to see them as a nuisance."

Salmon still cruise under the Golden Gate Bridge in their journey to and from the Sacramento to the sea. But California today is known for its tech boom, agricultural bounty, and urban chic—not salmon.

It is astonishing to realize what these California waters were like—and not that long ago. In a 1903 article in *Sunset Magazine*, fisherman J. Parker Whitney boasted of all the chinook salmon he could catch in *one day* at Monterey Bay, on an 11-ounce rod threaded with linen line. His enchantment with the bay and its salmon is poignant to read today, even hard to imagine: "The morning was warm and breezeless, and the glassy sea without a ripple . . . in these moderate swells were thousands upon thousands of salmon full of lusty strength and silvery glistening." He watched as the salmon pursued an immense mass of anchovies in the clear blue-green water, chasing the forage fish all the way up to the beach. By noon that day Whitney had caught fifteen chinook and by day's end at five o'clock, twelve more, for a total catch of twenty-five salmon, weighing a total of *482 pounds*. A photograph accompanying the story shows the fish hanging behind him. Gleaming and enormous, they are an apparition from another world, a portrait of loss.

Orcas were also in on the feast. Whitney writes that "one morning while fishing a large school of killers came around Point Lobos into the small bay which put a stop to all fishing for the day. I shall estimate the number at considerably more than 100. They came in like a pack of wolves, wild and frantic and leaping from the water like porpoises and plunging down after salmon and other fish . . . the killers

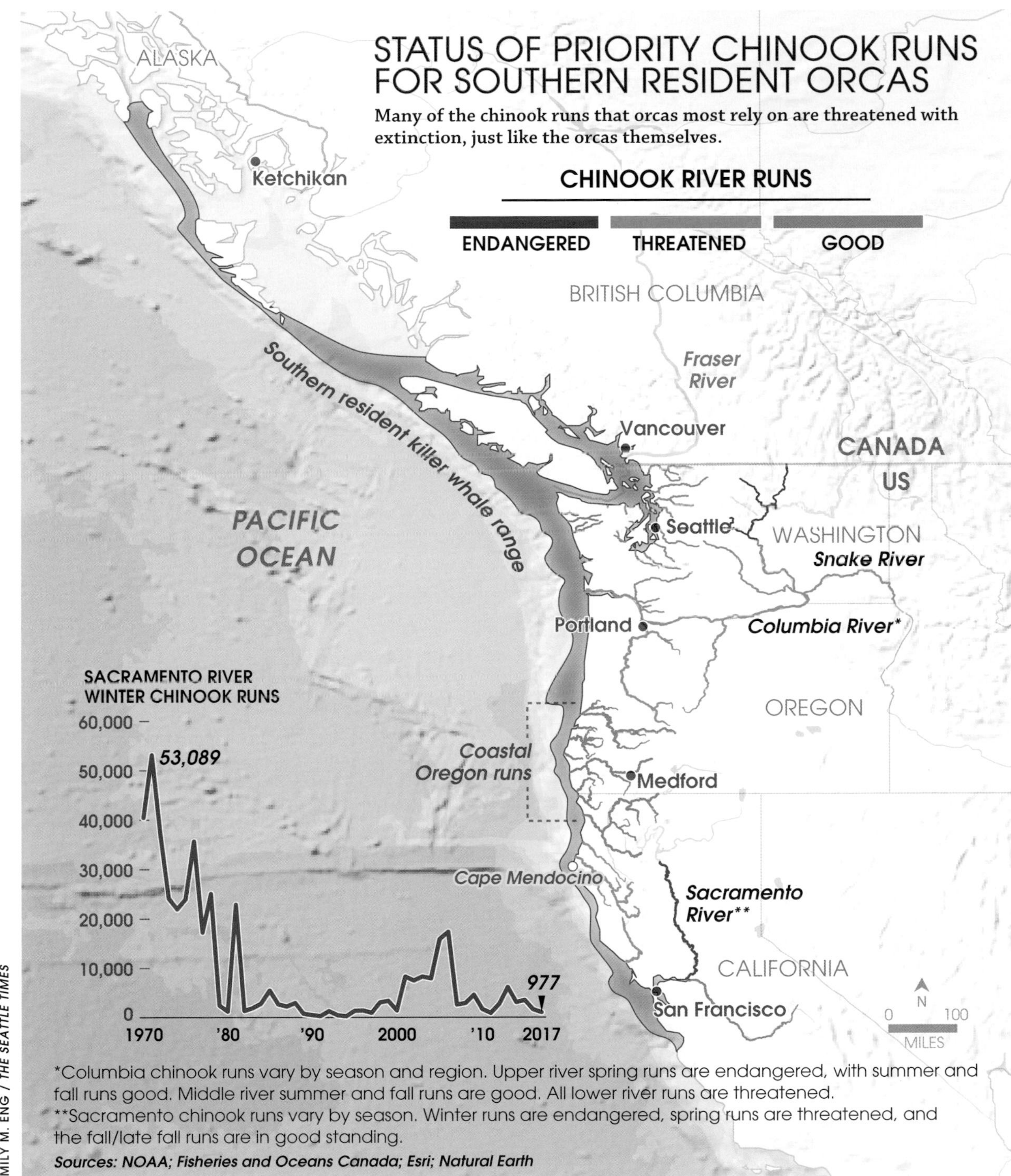

EMILY M. ENG / *THE SEATTLE TIMES*

TOP: Everything that hurts salmon happened first and worse in California. The fate of salmon there is a cautionary tale for the Northwest, the last stand for salmon and orcas in the Lower 48. This view is of San Francisco Bay and Foster City. *(Steve Ringman/*The Seattle Times*)*

BOTTOM: A US Fish and Wildlife fish culturalist grasps a thrashing Sacramento River winter-run chinook. Ripe with eggs, she is a rare find, and her eggs may be used for a captive population as a hedge against extinction. Southern resident killer whales still come all the way to California targeting these fish. But they are mostly chasing only a memory of abundance. *(Steve Ringman/*The Seattle Times*)*

although given as frequenters of the northern seas are plentiful in this coast and are the terrors of the sea. They come fearlessly about with their enormous dorsal fins projecting from four to five feet above the water, and slash about with their most powerful tails in a most threatening manner and I felt rather nervous at times as they came within short distance of my boat. Although they never have been known to attack boats or men." Maybe L25's mother was amid those black fins that so startled him.

I share this story because not that long ago the Sacramento was home to salmon runs second in the Lower 48 only to those on the Columbia River. What is happening in California is a cautionary tale for people of the Pacific Northwest: Lose enough for long enough, and pretty soon we could forget what we ever had too. Our birthright of abundance can be lost to the oblivion of an ever-receding baseline—and more quickly than we might think. Despite all the happy talk in Washington about salmon recovery, chinook are still declining, and the southern resident orcas are going down right along with them.

Battling Extinction on the Columbia and Snake Rivers

To experience a bit of the longest-running battle anywhere to save salmon, head to the Columbia and Snake Rivers. The salmon rescue and research blitz underway here is part of the largest fish and wildlife recovery program in the world, paid for by Northwest ratepayers in bills for their electricity, generated by the vast Columbia River Hydropower System. But after spending nearly *$18 billion* on fish and wildlife recovery in the Columbia River basin over more than two decades, not a single one of the thirteen runs of threatened salmon in the Columbia and Snake Rivers has recovered.

These Columbia and Snake River fish are critical to orca recovery, with multiple runs ranked high on the list of most important salmon runs for the southern resident orcas' diet. Tracking data by NOAA scientists and other research has shown the southern residents swimming back and forth between the mouth of the Columbia River and Grays Harbor, Washington, particularly in the early spring, to nail the glorious spring chinook—the fattiest of all salmon—returning to the river. But the cumulative effects of climate change, along with habitat damage including dams, hatcheries, and historic overfishing, have hammered these fish. The Lower Snake River dams, the latest built in the system, during the 1970s, have been a particular focus of both the controversy over Columbia and Snake River salmon declines and orca survival.

The historic abundance of salmon returning *annually* to the Columbia River before 1850 is estimated at from ten million to sixteen million fish—an incomprehensible wealth today, when even in the best years returns are a fraction of that bounty. Salmon also once traveled a far larger area. Historically, Columbia River chinook migrated 1,200 miles up the main stem of the Columbia River to Lake Windermere in British Columbia; 900 miles up the Snake River, the Columbia's largest tributary; and up to 900 miles south into the Snake's Nevada headwaters of the Owyhee River.

These epic journeys are now truncated by the dams built across rivers. More than 55 percent of the original spawning and rearing habitat for Columbia Basin salmon is inaccessible to salmon today because it has been walled off with impassable dams.

Until the dam-building era began in earnest in the early 1930s, the wild Columbia was a spectacular slasher of a river, sticklebacked with bone-crushing rock, swirling with sucking whirlpools, rainbowed with spray, and foaming with rapids, chutes, and drops as it smashed its way through rock walls in a 1,290-mile run from British Columbia to the Pacific.

But the river now is a unique source for hydropower, with a steep drop draining the snowmelt from mountain ranges across the 259,000-square-mile Columbia River basin. Eventually, 281 larger dams were built throughout the basin, home to 40 percent of the hydropower capacity of the entire nation and juicing powerlines all over the West.

But dams also alter rivers in many ways. The dams on the Columbia and Snake create a backwater effect that extends upstream, in larger impoundments of water. That also slows water-travel times between upstream and downstream points to seven to fifteen times longer on the Columbia and eight to thirteen times longer on the Lower Snake. That's tough on salmon, creatures of fast currents and cold water.

Hundreds of millions of dollars have been spent on dams on both the Columbia and the Lower Snake in an effort to improve fish passage and salmon survival. At Lower Granite Dam, the farthest inland, the cluster of slides, flumes, and buildings constructed over the years of salmon recovery projects makes the dam look like a water park in the middle of Washington wheat country.

As the spring migration gets underway each year, Lower Granite is an epicenter of salmon research and protection strategies. It's a dazzling and even bamboozling scene, as workers sort and ship juvenile salmon that just minutes before were minding their own business until encountering a migration massively altered by hydroelectric dams. Juvenile salmon headed downriver by the hundreds of thousands are tracked, counted, tagged, barged, and trucked past the dams. They are scooted past turbines in collectors and whooshed over the top of the dams on spilled water in an attempt to mimic the spring freshet. All of this is an effort to counter the basic problem that a soft, tiny fish no longer than a finger is trying to get past a cement wall 100 feet high and a battery of turbines grinding electricity from the current.

To track salmon survival through the dams, a portion of the juvenile migration downstream is intensively sampled and tagged. To see this in action is to confront the enormity of both the task and the effort.

In the spring of 2018, I watched as twenty seasonal workers crammed into trailers plunged needles into the bellies of tiny fish to insert tracking tags. More incoming fish shunted from the river for tracking and tagging zipped backward and uphill in translucent plastic piping to the taggers' workstation. The fish were temporarily zonked with a knockout potion, aptly called Aqui-S, to make them easier to handle. As the drug soaked in, the fish quieted, and workers standing in a line at a flume palmed each one to insert the tag, a little piece of glass about the size of a grain of rice, with a passive transponder in it. The tag enables managers to track the survival—or demise—of every salmon on its journey passing through detectors at each dam as it makes its way downriver from Lower Granite to Bonneville, more than 300 miles away, the last

This spiral fish flume is used to move juvenile salmon past Lower Granite Dam. Some fish will be loaded into a barge for transport to the sea, while others will be piped back into the Lower Snake River, the Columbia's largest tributary. *(Steve Ringman/ The Seattle Times)*

BRUCE M.
Juvenile Fish
Transportation

OPPOSITE, CLOCKWISE FROM TOP LEFT: At Lower Granite Dam, water thunders over the spillway to help sluice salmon to the sea; A US Army Corps of Engineers barge loaded with baby salmon chugs from Lower Granite Dam down the Lower Snake River; Fish are sucked up into a trailer where they will be microchipped to track their progress downstream through eight dams; Workers sort and tag juvenile salmon at Lower Granite Dam, injecting tags in the fish to track their survival through the Columbia River hydropower system. *(All photos by Steve Ringman/*The Seattle Times*)*

RIGHT: Baby salmon run a gauntlet of eight dams on the Columbia and Snake Rivers in their journey to the sea. *(Steve Ringman/*The Seattle Times*)*

dam on the main-stem Columbia River before the Pacific.

But in this industrialized river system, many salmon do not swim at all. Instead, they are taken out of the river and put on a barge to cruise toward the ocean. It's a strategy intended to spread the risk between salmon migrating in the river, and navigating the dams, and salmon getting a ride on a barge.

I once rode one of these fish barges run by the US Army Corps of Engineers, pulled by a tug downriver. I remember easing through the gigantic locks on the Lower Snake, the gates gradually opening to the river, the barge floating down to the level of the reservoir backed up behind the next dam. To feel that slow glissade and watch those doors magically swing open was to understand how totally the wild river has been supplanted by the river reworked by engineers. It's not really a river at all anymore, but a series of lakes regulated by locks and dams—what historian Richard White calls an organic machine. No wonder salmon have a tough time: masters of the Columbia's long-gone falls, chutes, and drops, they hadn't evolved for barging.

Dams affect not only juvenile salmon headed downstream but adults headed back home. Passage upstream through long slack-water reservoirs is perilous because all salmon, no matter where they live, are cold-water animals. The cumulative effects of climate change and multiple dams are pushing summer water temperatures into the seventies in some reservoirs for more than a month at a time. The longer salmon are in warmer water, the more susceptible they are to disease and dying before they reach their home gravel to spawn.

Orca L25 was about ten years old when the first dam, Bonneville, started churning out kilowatts in

1938. Today, for her family and for human families all over the region, the stakes on the Columbia and Snake are high. The dams are zero-carbon energy producers—critical as the region seeks to reduce its dependence on fossil fuels to fight global warming. Yet some orca experts maintain that the imperiled southern residents can't recover without removal of the four Lower Snake River dams.

Dam removal on the Lower Snake has been at the heart of a more than twenty-year court battle and political fight, with federal judges finding time and again that federal agencies are not measuring up to salmon-recovery requirements in operation of the federal hydropower system, despite habitat restoration, changes in flow regimes to benefit fish, and passage improvements at the dams.

Passage for juvenile salmon confronting the dams has greatly improved. But a poorly understood phenomenon called "delayed mortality" that kicks in after the last dam, thought to be linked to stress from the hydropower system, has made exact losses impossible to count.

This much is certain: historically, the Snake is believed to have been the Columbia basin's most productive river drainage for salmon and steelhead. It has supported more than 40 percent of all Columbia spring and summer chinook. But today these fish are at high risk of extinction. Some runs are already gone.

Despite more than two decades of federal judges—and many biologists and fish advocates—calling for a complete overhaul of dam operations on the Columbia and Snake to save salmon from extinction, that hasn't happened. Now another species—the southern resident orcas that depend on these salmon—also is in peril.

Seattle's Only River

Conditions for the salmon that feed the southern resident orcas are even worse in central Puget Sound. Seattle's only river, the Green-Duwamish, is a classic example of the losing battle between people and their wants and needs and the needs of salmon and orcas.

The Green River flows out of the Cascade Mountains north of Mount Rainier, first through state parkland, then farms, then ever denser development through the cities of Auburn, Kent, and Tukwila. There, the Green becomes the Duwamish River—an industrialized shipping channel serving the Port of Seattle, with access to Puget Sound and ultimately the Pacific through Elliott Bay.

The Green is still a top-priority salmon source for the southern resident orcas. But central Puget Sound also is a hub of intense human development and economic activity, an engine of commerce for the Pacific Northwest for more than a century. To paddle the lower Duwamish today is to witness all that can be done to a river without killing it. The scope of the alteration here of what once was prime salmon country is almost impossible to grasp: it is so hard to see or even imagine what is no longer there.

From a human perspective, the transformation was gradual. From the deep-time perspective of the salmon coming and going from this river over the last ten thousand years, or the orcas that depend on them, it all happened in the blink of an eye, with no chance to adapt.

In his elegiac book *The Price of Taming a River*, author Mike Sato tells of the more than one hundred years' war on this river and its watershed, the relentless remaking of this place of natural

TOP: The Duwamish River meets Puget Sound at the Port of Seattle. Massive alteration and degradation of the Puget Sound ecosystem is pushing the southern resident orcas to extinction. *(Steve Ringman/*The Seattle Times*)*

BOTTOM: Industrial development crowds the Lower Duwamish Waterway. The river has been straightened, dredged, and walled off from its floodplain and its natural estuary, and the river's wetlands and side channels have mostly been filled. Salmon that orcas rely on still travel through these waters, where a major Superfund cleanup and other restoration projects are underway. *(Steve Ringman/*The Seattle Times*)*

Visitors in the viewing area at the Ballard Locks fish ladder watch as coho and chinook salmon head home to freshwater to spawn. Too often there are no salmon to see here as chinook and sockeye decline. *(Ken Lambert/*The Seattle Times*)*

abundance, with dragline and steam shovel and suction dredge. What once was a wetland alive with clouds of ducks at the Duwamish delta is now Harbor Island, at 305 acres the largest man-made island in the world when it was built. It took eight years and 24 million cubic yards of silt to build Harbor Island in the early 1900s, using dredge spoil from leveling some of Seattle's original hills with water cannons—also done to create more buildable land.

The flood-prone and meandering lower Duwamish was next to go, straightened and dredged to create a waterway for industry. *Rectified,* as the saying was. Today the lower river is not a river at all but, rather, an industrial shipping channel with three turning basins, a deep-draft navigation channel, and levee walls. Forests along the banks were once tall enough to shade and cool the river and rain down a feast of bugs and leaf litter that stoked shredders and grazers, caddis flies, and mayflies that fed salmon. But by 1920, Sato reports, there were no forested banks, no shadowy pools; only one original bend of the river was left within the city limits of Seattle.

But these were not even the most dramatic changes to the pre-settler-era Green-Duwamish watershed. For that, look to the building of the Hiram M. Chittenden Locks and the Lake Washington Ship Canal. This was dramatic surgery of the kind confidently cut into the land and water with no thought to the violence done to nature's hydrology, the lives of native species, and the region's first people. Native Americans who didn't get out of the way were forcibly removed and their homes burned.

Even if their homes remained, the lands and waters and tidelands that had sustained native people for countless years were filled, diked, drained, and channelized. Forests along the rivers and in the uplands were cleared for development. That depleted the wood, berries, roots, medicines, plants, and animals that native people needed to survive. It was no different for the salmon, as their birth houses were excavated to oblivion or buried under fill.

Chittenden, of the US Army Corps of Engineers, oversaw the creation of two cuts, the Fremont Cut between Salmon Bay and Lake Union, and the Montlake Cut between Lake Union and Lake Washington. The locks at the west end of Salmon Bay, built to create access to Puget Sound, are used today mostly by recreational boaters, who make up more than 80 percent of the traffic through the locks.

The replumbing of the rivers of central Puget Sound that resulted from building the locks is so dramatic, it is hard to track without using historic maps to recall what used to exist. Lowering Lake Washington caused the Black River, formerly the outlet of the lake at its south end and the original route to the Cedar River for salmon through the lake, to dry up. Salmon that once found their way up the Black River to Lake Washington and the tributaries that flow into it, such as the Cedar River, have for a century been unnaturally rerouted through the ladders of the locks, north and west of their original migratory path. Then these salmon confront miles of concrete-sided, straightened navigation channels before arriving in a lake thronged with voracious predators. Their task is to find their tributary river and to move upstream to eventually—if they survive all that—spawn. It is not remotely the same journey.

Taken too were the lifeways of the Duwamish people and their relatives who had homes and fishing grounds all along the Black River. Duwamish descendant Joseph Moses described the day the Black River disappeared, August 28, 1916: "That was quite a day for the white people at least. The waters just went down, down, until our landing and canoes stood dry and there was no Black River at all. There were pools, of course, and the struggling fish trapped in them. People came from miles around, laughing and hollering and stuffing fish into gunny sacks."

But engineers, developers, farmers, and settlers still wanted more buildable, tillable land—tidy and reliably dry, stable places on which to settle and build. Yet gnarly big mountain rivers stormed into broad, flat valleys, blowing out whatever was in their way, avulsing entire floodplains. These rivers made and demolished gravel bars, felled trees on their banks with their chewing current, and created spectacular logjams. The jams split the rivers' waters and put them to work digging deep pools, heaping up gravel, and sneaking off into side channels—perfect salmon habitat. Constantly busy, the rivers sluiced down loads of silt and gravel and rock from the high mountains, rolling and tumbling even whole trees and boulders in big winter and spring flows. They fed and recharged miles of juicy marshes and wetlands and sloughs, springs and groundwater seeps.

These natural cycles of seasonal flooding made central Puget Sound a region of complex, braided, shifting, living river systems. This was a problem for people. Too much land in central Puget Sound was flood prone, wet, unworkable, unbuildable in too many places for too many months. The rivers wouldn't stay put. They were unreliable, untrustworthy. They did whatever they wanted, wherever and whenever they wanted to.

Imagine the work, without modern heavy equipment—or safety regulations—to clear these rivers' channels of large woody debris. When all else failed, farmers used dynamite to break up logjams and divert channels however they wished. They drained the wetlands, diked and filled the sloughs and marshes, encased the rivers in armor and levee walls. But even that wasn't enough. There was still flood after flood.

On the Green-Duwamish, first came the flood of 1906, as Sato recounts, a howler born of two weeks of rain, of course in November, with warm Chinook winds thrown in to melt the October snows in the high country. Before long, a torrent of floodwaters tore into the downriver towns as much as *24 feet* deep in some places. The Interurban streetcar line from Seattle to Renton and Auburn was cut off. Telegraph lines were downed, and logs and driftwood rammed and threatened bridges. People fled their homes, leaving their cows stranded on islands amid terrifying floodwater.

More bad floods in 1917, 1933, and then 1946 were the last straw. "More land for industry" was the battle cry of engineer Howard Hanson in a 1957 landmark article in *Pacific Northwest Quarterly*. He argued for building a dam on the Green River, to hold winter floodwaters, and for excavating and straightening the Duwamish all the way to Tukwila. Then, he promised, "old man river, from Elliott Bay to Auburn, will be under control."

Remaking the Duwamish

That's for sure. In the spring of 2018, I took river advocate David Shumate up on his invitation to meet him at his house on the lower Green River and experience the Green-Duwamish in a canoe. I arrived on a late summer day, with rain threatening. His house was a surprise in this dense urban area, tucked back amid mature cedar trees.

We edged carefully into his canoe, found our most comfortable balance, and pushed off. I felt it all over again, instantly, that wonderful silken glide and feeling of freedom that comes as our terrestrial life gives way to the seduction of water, whether fresh or salt. No matter the boat—grand or small, whether motorized or sail-powered or paddled—there is something magical in those first waterborne moments. No longer a clumsy two-legged, you're afloat.

As we headed downstream, we caught a glide on the current. Here, the river was so much lower than the tops of the earthen levee walls, there was nothing much to see along the riverbanks, smothered in blackberry brambles. We had no sense of riverside landmarks as the walls passed by far overhead in their sameness. As we drifted, I learned the many ways people have, over the years, contrived to armor riverbanks. Here was a veritable library of techniques: Tires, arranged in stacked rows and rammed into the mud banks. Mounded earthen levees, towering well overhead. Rocks piled, heaped, and stacked. Cement walls. Steel sheet piling, creosoted railroad ties, boards. They all fenced the river firmly in place.

A natural bank or entrance to a side channel was rare. But wherever we found these, so had the birds. Sandpipers fluttered over soft sand. A heron stood

A mountain of scrap metal sits at the edge of the Lower Duwamish, long ago converted to an industrial shipping channel and waterway. *(Steve Ringman/*The Seattle Times*)*

in shallows, ready to spear a meal, its elegance a contrast to the leering insurance-company mascot gecko on the billboard on the banks overhead. We passed an abandoned homeless encampment deep in the tangled underbrush of the banks, where sweatshirts, socks, and a wet sleeping bag left behind were on their way to falling into the river. A rent-a-bike glowed green where it had been thrown into the water, and a kingfisher chattered overhead. Yet for all the human encroachment, the river was still its feral self; it smelled rankly of mud, and I could see its green flow, diatoms thickening the water.

That changed as we dropped lower, into the Duwamish Waterway. There, the human intrusion was nearly total, startling in its supremacy and ubiquity. Aircraft lumbered low over the water, headed to the runway at Boeing Field. The tail fins of planes jutted over the water, parked on acres of asphalt paved right to the bank. The din was continual: banging, crashing, backup alarms, traffic, the roar of cement plants and heavy equipment. We floated past a front-end loader pawing at heaps of scrap metal piled high right along the river, tipping our heads back in amazement at stacks of crushed cars crammed along the east bank. Staged along the wall of the opposite shore were barges loaded with steel shipping containers in multiple colors bright as building blocks. One barge was loaded five containers high, with a tour bus parked on top. The bus was emblazoned with a leaping orca.

To be sure, a lot of work has been done and is underway by governments, businesses, nonprofits, and volunteers to restore the Green-Duwamish to something more like health. A $342 million seventeen-year Superfund cleanup is proceeding in the lower Duwamish, with the goal of removing and remediating decades of toxic pollution. The Port of Seattle has completed 93 acres of habitat restoration in the watershed and Elliott Bay, with plans for more. Community shoreline access and habitat renewal have brought some measure of life back to the Duwamish. All sorts of new directions and experiments are being pursued to improve water quality, and revegetation and reforestation are underway. Even engineered, floating wetlands—called bio-barges—are an experiment, utilizing rafts planted with wetland species to provide a bit of the ecological function lost along with all that used to be here.

The Duwamish has been paved and poisoned under practices mostly disallowed today. "For so many years the river was just used as an open sewer," Shumate said as we paddled. Those so-called point sources of industrial pollution have been greatly reduced, but much harder work remains. An insidious dispersed seep of toxics and pollutants is reaching rivers and streams and the Sound by actions mostly legal—and ordinary: from driving and parking cars to washing them in the driveway; using pesticides and herbicides on gardens, lawns, and golf courses, as well as harsh chemicals and solvents in homes; improperly disposing of pet waste, left behind instead of picked up and put in the trash.

There is still illegal dumping, of course, and violations by industry sometimes bring fines from regulators. Government agencies pollute, too, dumping raw sewage in rivers and the Sound due to various malfunctions or because a sewer system overflows in a big rain. Such bacterial bombs are relatively infrequent, and the region is spending billions to nearly eliminate them once and for all—at least that's the plan. Yet that doesn't do anything for the pollution carried in untreated storm runoff that reaches the rivers and streams, Puget Sound and its bays, every time it rains.

It was raining now, pocking the river, stoking discharge of the wet-weather runoff from all the surrounding roofs, parking lots, roads, and industrial facilities and junkyards that flows untreated into the Duwamish. Rain runoff carries oil, grease, dirt, tire dust, soot from engine exhaust, and a stew of pollutants from any and every hard surface. The biggest source of pollution to Puget Sound, rain runoff carries a mix that is so toxic to coho salmon—food for southern resident orcas from late summer on—that many female fish full of eggs die before they can spawn in urban creeks.

Our urban waters also take the brunt of our personal waste stream. Sewage treatment plants don't remove the plastic fibers rinsed into laundry water from synthetic clothing, including nylon, acrylic, and polyester outdoor gear. Pharmaceuticals that we pass through our bodies also are loaded into wastewater. Puget Sound salmon are on drugs—Prozac, Advil, Benadryl, Lipitor, even cocaine. Those drugs and dozens of others are showing up in the tissues of juvenile chinook, researchers have found, thanks to tainted discharge of our everyday sewage.

The estuary waters near the outfalls of sewage-treatment plants, as well as effluent sampled at the plants, are cocktails of eighty-one drugs and personal-care products, along with nicotine and caffeine, at detected levels that are among the highest in the nation, one experiment showed. The medicine chest of common drugs also included Flonase, Aleve, Tylenol, Paxil, Valium, Zoloft, Tagamet, OxyContin, Darvon, fungicides, antiseptics,

The Howard A. Hanson Dam, built in 1962, blocks fish passage to half the Green River watershed, including nearly all its best salmon habitat. Providing fish passage would open 100 miles of spawning habitat for chinook, coho, and steelhead. *(Steve Ringman/*The Seattle Times*)*

and anticoagulants. Not to mention ciprofloxacin and other antibiotics galore, according to Jim Meador, an environmental toxicologist at NOAA's Northwest Fisheries Science Center in Seattle. He was lead author on a 2016 paper published in the journal *Environmental Pollution* that revealed just what we flush into the Sound. The amount of drugs and chemicals from all wastewater-treatment plants dumped into the waters we live near could be as much as 97,000 pounds every year, the study found, from 106 publicly owned facilities discharging to Puget Sound.

An article I wrote about the study was maybe the only time my reporting has wound up on CBS, on *The Late Show* with Stephen Colbert in a routine in which Sammy the Salmon is so hopped up on drugs he requires, once going into withdrawal, a splashing of water from Puget Sound to get his fix. It was funny, no doubt. But to see Stephen Colbert on national television holding up my story and his drug-crazed salmon puppet was also an arrow to the heart. Seattle's salmon habitat had become a national laugh line.

Green River's Farm Fields Turned to Warehouses

In the late summer of 2018, I took a trip to see the lower and middle Green River. The vegetable and berry farms, pastures, and dairies that were once here on some of the best soil in the state are today

The Green River, walled off by levees, passes through industrial and retail zones of Kent between Interstate 5 and the West Valley Highway. *(Steve Ringman/* The Seattle Times*)*

mostly the stuff of historical photos. What was a cornucopia for Seattle is now the second-largest warehousing and distribution center on the West Coast. To protect the buildings, on stretches of the Green River through Auburn and Tukwila every tall tree was cut down to construct armored levees, some with rock walls built to launch boulders into the river in high water, to fill any pools the river might get a notion to form. Big pools, the kind chinook crave.

To erect the Briscoe-Desimone levee to protect portions of Kent, Renton, and Tukwila from flooding, engineers chopped down an 80-foot cottonwood and drove steel sheet piles into the riverbanks, then faced them with rock—a design selected in 2013 for maximum speed of construction at minimum cost and taking as little land as possible from parking and low-rise office parks. "This is basically a ditch," said Katie Beaver, a river steward for King County, leading my tour that day.

We had started out higher in the watershed, in the middle Green at the Green River Gorge, near Black Diamond, driving country roads and stopping along the shore where the river was still itself for long stretches. But as we dropped lower in the watershed, to the urbanized lower Green through what is now a booming warehouse district, the transition to where we stood at the Briscoe-Desimone levee wall backed by an asphalt parking lot, was emphatic.

Beaver's job, among others, is to work with planners to identify places along the river where walls and levees could be set much farther back from the river, to give the river room to move and to improve salmon habitat, a strategy with which the county and its local flood district—legally separate entities run by the same council members—has had some success. But so much of the river's shoreline is already hardened. Though more than 80 percent of the Green is behind walls, mayors and businesses and farmers want to build even more of them. The fight for what's left of this river for salmon—and orcas—is still underway.

Nearly every stretch of the river and its floodplain needed for salmon is already taken by people, contested and fought over: whether to keep a golf course or a community garden or take them out, to give the land back to the floodplain as open space; whether to build even more commercial development, warehouses, and apartments. People already living along the banks also treasure their views, cutting the trees to put their lawn furniture atop levee banks, to watch the river slide by between walls.

With the lower watershed already so intensely built and developed for housing and commercial and industrial use, what has happened in the upper watershed is all the more poignant. Here the bobcats and cougars and bears and deer still outnumber the people. One reason is that in 1906 the city of Tacoma built a dam to divert the river's flow for drinking water and the thirsty pulp industry, shutting this upper watershed off to development. Today, the upper watershed is mostly publicly owned by various agencies. The land has been heavily logged but is managed under long-term habitat agreements, so its condition should only improve. But salmon are fenced out of more than half of this watershed—and all of the best of it.

In the upper Green, the Howard A. Hanson Dam was completed in 1962 in part to manage flooding, particularly in the heavily developed lower reach of the river in places including Auburn, Kent, and Tukwila. Built without fish passage, this federal dam walls off more than 100 miles of river and side channels upstream—spawning habitat just waiting for chinook, chum, coho, and steelhead. But despite decades of talk and more than $100 million spent, there still is no fish passage at the Howard Hanson.

All parties agree this dam operated by the Corps of Engineers should provide fish passage. "The only controversy is why it is taking so long," said Fred Goetz, Endangered Species Act coordinator for the Corps' Seattle district office. The Corps received congressional funding for a juvenile fish-passage facility here in 1999 and began construction in 2002. But the agency quickly busted the project budget. With $108 million spent, the Corps stopped work in 2011. By now work has been stalled for so long that the process to design, approve, and fund the work must be restarted. NOAA has set a deadline for operational fish passage at the dam—in 2031.

"This may be the most important single project that can be done for salmon recovery in Puget Sound. We have a tremendous opportunity here," Goetz said. "The continuous decline of the orca is not helped if we are not able to accomplish these big missions."

I hiked the upper watershed in the fall of 2018, just to explore what's up there, where the fish can't go. It's beautiful country. There are big cottonwoods, fir, and spruce that shade the river, winding through tumbled woody debris, pools, and side channels. It was easy to see how the Green River gets its name, purling over the logjams created both by nature and by river engineers. The Tacoma utility and the Corps

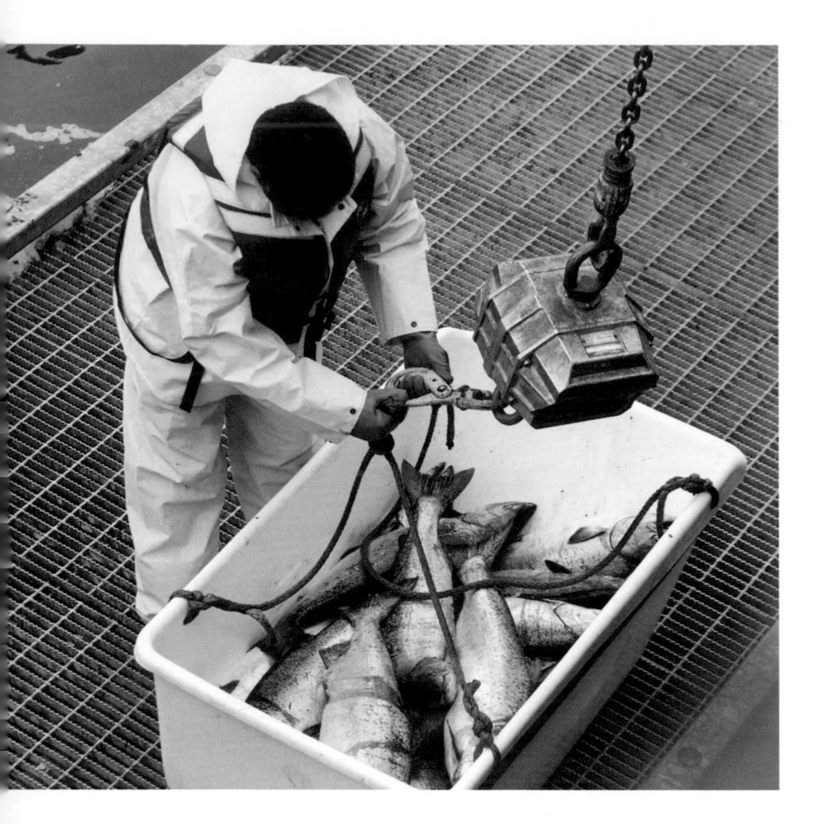

Chinook runs have been so poor that Muckleshoot tribal members, who have treaty-protected fishing rights in Elliott Bay and the Duwamish River, have been limited to as little as twelve hours of fishing in some seasons. *(Steve Ringman/*The Seattle Times*)*

have been investing in habitat enrichments for the big fish, when they finally get here someday. It's a palace for salmon, with no king.

Survivors

Yet life stubbornly persists. Even in the thick of the industrial reaches of the Duwamish, eagles flap by, chasing a seagull. And people still come here to fish—just as they always have.

On a cold, foggy morning in the summer of 2019, I made my way to a boat ramp, hard by homeless encampments and under the roaring traffic of a highway overpass in Seattle, to watch and talk with Muckleshoot tribal members bringing in their catch. Some ninety boats turned out for what was only a twelve-hour opening—all the river could support for this treaty-protected fishery that year. But there they were: big, beautiful chinook, still silver bright from the sea. In this setting, they seemed as unlikely and out of time as dinosaurs. And yet these people, and these fish, the river's first inhabitants, were still here.

The Muckleshoot Indian Tribe has invested heavily in the Green-Duwamish River, and their treaty entitles them to half the salmon catch. But half of next to nothing is just about nothing. The catch was so small in this briefest of openings that there would not be another fishery allowed that season, despite the treaty's promise.

These losses were not so much intentional as not thought about. They didn't matter to the settlers who remade this place. Historian Coll Thrush at the University of British Columbia, in his superb article "City of the Changers" in *Pacific Historical Review*, notes the remarkable lack of concern as newcomers blew up, carved up, replumbed, and remade the lands and waters where they had only recently arrived.

"The people who did all this thought they were improving on nature; they thought they were making it better and more efficient," Thrush told me. "They really thought they were doing the right thing, and they were completely dismissive of the effects on the salmon and indigenous people, which are really the same story. You had indigenous people starving within sight of the Smith Tower [in Seattle]; there were no fish."

Thrush grew up in Auburn and knows its landscape. He sees an unreconciled history held in this land and its people. "I feel like Seattle and Puget Sound is still coming to terms with its own past and the consequences of that past," Thrush said. "We

UNDERSTANDING SALMON-FRIENDLY SHORELINES

Land use and restoration decisions made across Puget Sound's waterways will determine whether salmon in our rivers will recover or continue to decline. Understanding the elements of healthy shorelines is key to this issue.

SHORELINE HABITAT

Natural bluffs feed beaches. Forage fish lay their eggs on beaches while young salmon and birds use beaches and shallow waters for feeding on insects and other prey.

A hardened shoreline lacks shrubs and trees, deflects waves, narrows the beach, and erodes it. Biodiversity, including crucial habitat for forage fish, shorebirds, and baby salmon is lost.

RIPARIAN ZONE

A healthy stream is shaded by native plants, cooling the water. The shoreline meanders through its floodplain as water levels change seasonally. Gravel provides material for salmon to dig their nests.

An unhealthy stream has no shade, raising water temperatures that can kill salmon. It is channelized and cut off from its floodplain. There is no insect rain or leaf litter to provide a food source for a diversity of species and no gravel for salmon to dig their nests.

LEVEE TYPES

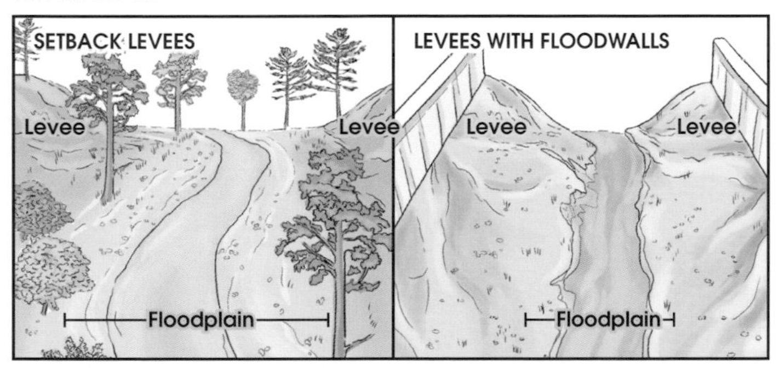

Setback levees create room for a river to move, managing floodwater while retaining a natural shoreline. Larger setbacks can allow public access and create space for a greenbelt to shade water and keep it cool for salmon.

Levees with floodwalls require less property to build, but they offer fewer restoration opportunities.

RAINWATER RUNOFF

FORESTED RUNOFF

URBAN RUNOFF

In a forested landscape, vegetation and soil hold runoff, cleansing the water and moderating flows.

Rain runs off hard surfaces, carrying untreated pollutants and creating surges of water into streams that damage habitat. Green stormwater infrastructure, including home rain gardens, can help moderate stormwater.

Source: King County Department of Natural Resources and Parks

EMILY M. ENG / *THE SEATTLE TIMES*

are all inside that story; we are all part of it—we have a right to speak of it and a responsibility to speak of it."

These lands and waters also speak for themselves, in layers of history and time, read in pollen records and shell middens, in extirpation, in displacement and persistence. And even in the memories of matriarch orcas like L25, still here through more than eight decades.

"The orcas—they know; they have seen it all," Thrush said. "They haven't seen the dam on Eagle Gorge on the Green, but they have seen its effects; we are past the point of thinking they are 'just animals' without consciousness. What might the story look like from their perspective?" Long-lived, the southern resident orcas are repositories of memory and experience, and they have witnessed so much change in less than two generations. "They have their own histories that they carry," Thrush said. "They are an archive."

A Race against Time

It is dizzying to think of the changes one extended family of orcas has been forced to contend with. L25 was born in about 1928, before the first dam on the Columbia, Rock Island, was built, and Bonneville Dam was still another decade away. There was no Hanford Nuclear Reservation, no Grand Coulee Dam; there were no dams on the Lower Snake River. The state's population was about 1.5 million people.

Over L25's lifetime, the state's population has grown to nearly 8 million, with no end of growth in sight. As the region booms, salmon and orcas are in a race against time. Under siege since the settlers arrived with their draglines and steam shovels and suction-pump dredges, the threats to the southern residents' home waters have metastasized.

Beyond central Puget Sound, where the destruction started, are some of the Puget Sound region's fastest-developing landscapes: the suburbs. More growth means more people and, depending on how growth is managed, more pollution and more runoff, as forests and open spaces that absorb the heavy rains and filter pollution are paved over or converted to housing, shopping centers, office parks, and all the rest. With the urban core already built out and devoured, what is happening here now is destroying the best of what is left.

Regional housing shortages and skyrocketing housing costs make the problem worse. The burgeoning job centers in Seattle and the east side of Lake Washington have brought more and more people chasing six-figure tech work. That has pushed people to places like Granite Falls, 42 miles north of Seattle in Snohomish County. This city, near the Stillaguamish River, notched the second-fastest rate of growth among cities in the four-county central Puget Sound region of Snohomish, King, Pierce, and Kitsap Counties in 2018–2019. One reason is affordability. The price per square foot of a home that year in Granite Falls was less than half that in Seattle.

I went to go see what money buys for people flocking to the suburbs to enjoy a lifestyle with more space for less money. "Sold out of inventory, more coming soon!" read the sign stuck in the ground at Suncrest Farms, a new housing development in Granite Falls. Swing sets and backyards beckoned.

Just across the street was a pasture, where a lone horse up to its belly in spring daisies swished its tail. How much longer, I wondered, watching

The Stillaguamish Tribe bought this farmland along the Stillaguamish River to restore the floodplain and improve conditions for young fish. The tribe is buying land wherever it can for fish habitat. *(Steve Ringman/*The Seattle Times*)*

that horse, would this open field, part of a classic old farm with its meadow transitioning to tall firs, be there? How long before it, too, would become a subdivision? In these parts, a happy horse in a daisy-flecked meadow looked as endangered to me as a Puget Sound chinook or southern resident orca.

Just down the road from these burgeoning housing developments was the river—and in its south fork a last ditch against extinction for fall chinook: a captive brood facility, run by the Stillaguamish Tribe. Here, salmon are raised in tiny hotels hung on a wall, one salmon to each water-filled plastic box. As they grow, the fish progress to living together in large circular tanks. The biggest adults circled endlessly in their tank, kept in half light. Just like the captive brood of winter chinook I had observed in California, they will never know the sea.

Facilities like these are expensive—and a sign of desperation. In both cases, the captive brood facilities are being used to preserve the genes of a run of chinook down to just a few thousand fish. Tribal Chairman Shawn Yanity isn't happy his tribe can no longer fish for this population of chinook in the Stillaguamish. He is saddened that turkey and ham are on tribal tables, even for important ceremonies: fish for the first salmon ceremony is purchased from neighboring tribes. It's not supposed to be like this; it never was like this.

But where once there was an ancient alliance between people and salmon, Yanity said, today people are in a contest with salmon for the last of what's left. What's at stake is identity, culture, and abundance not found in any store. Yanity wants his people fishing again. "It's the teachings, the stories the elders tell, the protocol and preparation for fishing and hunting," Yanity said. The captive brood is both necessary and crushing. "I don't want to see my culture in a tank."

This attachment to salmon isn't only a Native American thing. In Washington, salmon are still a secular sacrament, what many people of all creeds and races say are big part of what makes Washington home. Talk to sport fishermen or old duffers living back along the deep holes of the Skagit River, hoping for a slab-sided chinook to put in the smoker. Even people who don't fish just like to know the salmon are still here. It's what these fish stand for: a still-functioning natural environment, the Pacific Northwest a lot of people mean when they say they live here—a place not just like everywhere else.

That's both good and bad news for Jeff Davis, director of conservation for the Washington Department of Fish and Wildlife. He knows people here still care about salmon. But he is also well aware he is losing this battle. A congressional tour of suburbs south of the state capital in Olympia, where pavement is pressing deeper into the remaining bastions for salmon, left him in despair. Restoration work is underway all over the state, but it is being outpaced by habitat loss. He urges a new understanding of how we use land and water that lives up to what we say we love and money to fix what we have already wrecked.

"We haven't gone far enough," said Ron Warren, the director of fish programs for the Washington Department of Fish and Wildlife. I had called him and other fish managers to ask them to take stock of where we stand. "We have a no net loss of habitat policy, and that doesn't seem to be working," Warren said. "We have to somehow change that paradigm; there has to be a gain. We have to assure ourselves we are going to get something for the continued creep into habitat that lessens the likelihood we will ever get fish. Otherwise we will never have the cool, clean water that fish need. You have to start telling people what you think and be truthful.

"I am not knocking the choices society has made, but we have to challenge ourselves and decide, Do we keep doing that? And if we don't, how do we pay for that? How do we get four million more people here, and where is the planning for that—the new treatment plants, the runoff from the roads and that many cars? These are the things we need to look at."

Downtown Orcas

The southern resident orcas still seek the fish returning to Puget Sound rivers, surging even all the way into the urban waters offshore of downtown Seattle, hunting chum, coho, and chinook. The special time for Seattle-area residents is when the southern residents, in their final seasonal rounds of the year, come here at last. Downtown orcas. Who else has that?

Sometimes the southern residents are here for days on end, thrilling ferry riders crisscrossing central Puget Sound and people flocking to beaches all over West Seattle and Vashon and Maury Islands to watch orcas blow and breach, right offshore. One day in November 2018, J, K, and L pods were all here at once. Dozens of orcas were cartwheeling and spy-hopping, right past the Superfund site of the Asarco Smelter at Ruston near Tacoma, right past the dense-packed housing along the long-ago logged-off hillsides. They sculled underwater upside down, their bellies glowing white through the green water, and slapped their pectoral fins and flukes

Stillaguamish chinook circle in a captive brood to protect the genes of fish so scarce they could be the last of their kind. *(Steve Ringman/*The Seattle Times*)*

A couple practices fly casting skills in the middle of the Green River at Flaming Geyser State Park near Black Diamond. The Green River remains one of the most important sources of salmon for southern resident killer whales. *(Lindsey Wasson/* The Seattle Times*)*

seemingly just for fun or maybe simply to hear the loud, resonant, smacking sound.

As the sunset painted the water gold, people thronged the beaches and shorelines, enchanted all over again at what it means to live here, in a place still alive with salmon and orcas on the hunt.

The southern resident orcas that roam the region's urban waters are like black-and-white-robed judges of a truth-and-reconciliation commission. They remind us all of what is still here and what is at risk—what we took for ourselves, what we took from them. Their hunger is an indictment, because we caused it. Their hunger points to what's missing. At stake as the region gets richer is whether it also will get poorer, with only the grandmother orcas remembering the salmon that used to be.

To better understand what the southern resident orcas need, I traveled north, to lands and waters where orcas are still thriving. The land of the northern residents is a place that looks a lot more like Puget Sound used to. The orcas living there are the same animal as their urban counterpart, the southern residents. The only difference between the northern resident orcas that are thriving and the southern resident orcas headed to extinction is . . . the millions of us who live in their home territory. They were here first. They belong here. But now, they contend for survival amid all of us.

CHAPTER SIX

A Land to the North

Hundreds of dolphins flew over the sea's glittering surface. Suddenly a black fin showed, rising from the water, bigger . . . and bigger . . . finally towering nearly 6 feet high. Soon black fins surrounded the bow of the *Wishart* as Oceans Initiative scientist Rob Williams piloted his research boat through a scatter of remote islands northeast of Vancouver Island, British Columbia.

The environment of the northern resident orcas north and east of Vancouver Island is more like Puget Sound used to be, and not that long ago. *(Steve Ringman/*The Seattle Times*)*

Ernest Alfred of the Namgis First Nation sprang up from his seat in the cabin and hurried to the bow. He began to sing a welcome song to the orcas, low and sweet, beating time on the boat with one hand. Alfred's family crest is an orca, and his family name comes from the orca clan. "We go back to the time when we all used to speak the same language," Alfred said. "Some of those whales may even be old enough to remember hearing me sing that as a boy."

We started out at Alert Bay, on Cormorant Island, to pick up Alfred and begin our journey to OrcaLab on Hanson Island, just to the south and east. Williams steered the boat on, with scientist Erin Ashe, executive director of Oceans Initiative, navigating. They were here on a research trip for Ashe's long-running ecological study of Pacific white-sided dolphins. They make fieldwork a family affair, taking along their daughter, Clara, age five at the time. The *Wishart* is named after their first dog, who taught himself to sniff out dolphins for Ashe. The late great Wishart the dog is no more—but now Molly, a solid, athletic Chesapeake Bay retriever, was curled up on the boat cushions in the cabin, keeping a devoted eye on Clara.

Williams had helped introduce me to Paul Spong, one of the world's foremost experts on northern resident orcas, and we were headed to Spong's remote research outpost at OrcaLab. I had just arrived from Seattle, and encountering orcas without even trying to, within minutes of coming aboard, was a surprise, even a shock. Working with researchers in the summer habitat of the southern resident orcas in the San Juan Islands, it usually took anxious days of trying, waiting, hoping and elaborate logistics to do any work involving the southern residents. I'd grown so used to an atmosphere of catastrophe, absence, and loss amid the southern residents that to so easily just happen upon this healthy, romping pod of northern resident orcas in our very first hours together on the water was a restorative corrective, an uplifting reminder of what normal looks like.

We saw orcas everywhere we went. As the sun burned off the morning fog, the sea was alive with sounds. Sea ducks ran over the water, the air loud with their wings. Pods of dolphins stampeded in arching dives so fast they churned the water. The northern residents live not just in a different place but in another world.

Northern resident orcas enjoy waters that mostly are cleaner and much less affected by boat and ship noise than the urban waters and industrial shipping lanes in much of the southern residents' home range. The northern residents also have more abundant salmon and a greater variety of salmon populations on which to feed.

The northern and southern resident orcas share some of the same territory and have similar family bonds and culture. But while the southern residents are struggling to survive, the northern residents' numbers have more than doubled since scientists started annual population surveys in 1973. Because they are the same animal, but doing so well, the northern residents function for scientists as a kind of control population—a perfect point of contrast by which to better understand what is ailing the southern residents.

Scientists attribute the northern residents' population growth to several factors, beginning with a reduction in human insult, with the initiation of government protection from hunting marine mammals on both sides of the border. "The north-

TOP: Northern resident orcas cruise their home waters in Johnstone Strait, on the east side of Vancouver Island. The population of northern residents has been steadily increasing for decades. *(Dr. Andrew Trites/University of British Columbia)*

BOTTOM: The northern resident orca population has been growing steadily, showing what is possible for orcas when they have cleaner, quieter waters and more abundant and a greater variety of salmon on which to feed. *(Steve Ringman/*The Seattle Times*)*

ern resident population was persecuted quite a bit prior to the early 1970s," said John K. B. Ford, the pioneering Canadian researcher on the culture and ecology of resident orcas. "There was a lot of directed shooting and mortality, I think more than we understood."

Northern resident orcas also suffered in the live-capture era, though not nearly as much as the southern residents, which were easier pickings in the sheltered, shallower coves and bays of Puget Sound. Once the live captures stopped in 1976, both populations began rebuilding—although the

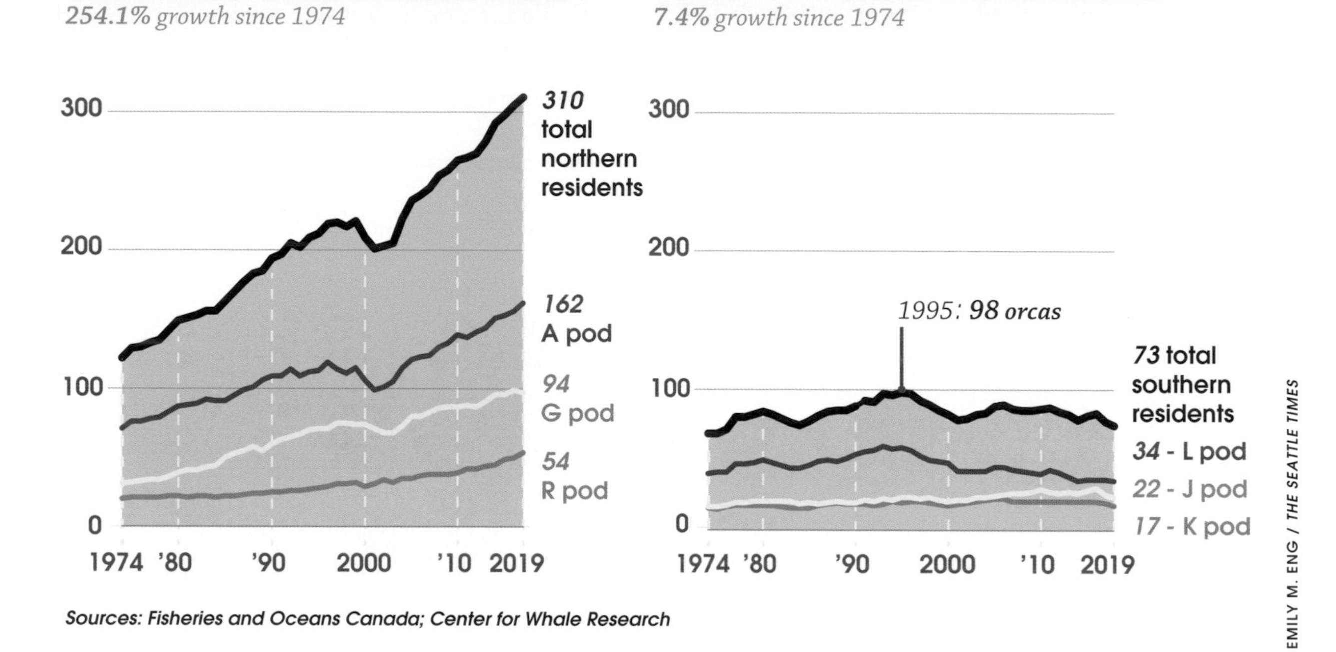

southern residents, with an entire generation of young orcas taken, did so more slowly. Southern resident orca family demographics were badly skewed, with so many young animals taken, Ford said. "That really inhibited the southern residents for some time. But both [populations] were generally increasing when there was sufficient food available and they were not being shot anymore."

It wasn't just the total number of orcas that died in the captures or, later, in captivity that mattered. Resident orcas are social animals that rely on relationships within and among family groups for survival. The captors took animals with important social roles and connections. Taking them from their families was particularly harmful, according to a 2007 paper, because it fragmented the social network crucial to the survival of family members left behind.

Both the northern and southern resident populations also suffered during a steep crash in chinook populations that began in the 1990s and lasted for five years. But after that, southern residents continued to struggle while the northern residents' numbers took off. The northern resident population has grown by a third—one hundred animals—since the early 2000s, to more than three hundred orcas, Ford notes, while the southern residents today are at their lowest population in more than forty years.

Food is part of the reason. The northern residents benefit from a wider variety of fish to prey on, including Washington-origin salmon from the Columbia River and Puget Sound that mostly head northward when they reach the outer coast. The northern residents—and fishermen working in Southeast Alaska ocean fisheries and off Vancouver Island—get first crack at those adult fish, before the southern residents. Researchers using underwater listening devices at Swiftsure Bank, at the entrance to the Strait of Juan de Fuca, have been surprised to realize how much time the northern residents spend in the southern residents' home waters—feeding on chinook.

To be sure, the northern residents are still classified as a threatened population by the Canadian government. It's still a small population, and it increases slowly. But Canadian scientists have recorded only one dip in the population since 2001, from 303 northern resident orcas in 2017 to 302 orcas in 2019. That slight drop has caused scientists to wonder if the environment has reached the limit of its carrying capacity for northern resident orcas. But scientists have been fooled by the northern residents before, when a drop in their population was followed by another surge of growth. Indeed, the best population estimate for 2019 was 310 northern resident orcas, the highest since 1973.

OrcaLab

It wasn't long before Williams slowed the *Wishart* for the approach to Hanson Island. Paul Spong was waiting for us onshore as we neared OrcaLab. He and Williams are old friends, and they waved to each other as Williams closed the distance. Williams tied off to a mooring buoy as Spong hopped into the *June Cove*, an old workboat with a plywood floor. He brought it alongside so Rob didn't have to risk the fiberglass *Wishart* on the rocks in shallower water.

We clambered aboard for the short crossing over to OrcaLab, saving pleasantries as Spong put on a pair of chain-saw ear protectors and throttled up the *June Cove*. Its monster engine shuddered the boat's metal frame as we rattled across the clear jade water. We hopped off and I caught my breath, trying to take it all in.

OrcaLab grows on the shore of the island like a huckleberry bush sprawling from an old-growth stump: organic and rooted into its place. Since 1970 Spong has created what has grown to a collection of simple low buildings on the island's shore. Even the door pulls are handmade from driftwood gleaned off the beach.

Spong and his wife, Helena Symonds, live here year-round, amid a bustle of researchers working with them to compile a vast archive of orca sounds from hydrophones. The underwater listening devices are deployed throughout the northern residents' waters, just offshore, and hooked up to speakers inside the lab and the house, where the sounds and windows everywhere toward the water create a sense of immersion in the orcas' world.

Spong and I went out on the deck to talk, leaning on a rail made from a branch. He had the calm ways of a man who has lived in remote country for a long time, studying an animal mainly by listening to it. He spoke just above a whisper, all that was needed in the quiet that enveloped us at the water's edge. The sky was oyster gray, the water a vitreous deep green. A humpback breathed in the distance as Spong, now in his seventies, talked about what brought and has kept him here on Hanson Island for some forty years.

"Essentially, we are a land-based whale research station," Spong said. "The reason we are land based is, we are interested in the lives of the orcas but we don't want to interfere with them. So we use remote systems we monitor and control from our lab. The

LEFT: Paul Spong founded OrcaLab in 1970 to document the daily lives—and spectacular sounds—of the northern resident orcas. He was among the first to appreciate orca intelligence, among the keenest in the animal world. *(Steve Ringman/*The Seattle Times*)*

OPPOSITE: Volunteers on their tiny observation platform monitor the northern residents from the most remote outpost of OrcaLab, on Cracroft Point above Johnstone Strait. *(Steve Ringman/*The Seattle Times*)*

first part of what we do is listening. We have a network of hydrophones that covers 50 square kilometers [about 31 square miles] in the core habitat of the northern residents."

The listening network is on twenty-four hours a day, recording the orcas. Because the orcas are so vocal, listening is a good way to begin to understand the comings and goings of their families, Spong said. "If it is daytime, we add a visual dimension, with a network of video cameras that enables us to see who is there and what they are doing." Spong's hope is to give the world a look into the lives of orcas, to create a better understanding of the orcas and their needs as a community, with families and a living culture deeply embedded in this place.

The work has become his life. "At first we were just visiting scientists," Spong said. "It became home." Over the years, he has felt profoundly inspired and instructed, even mentored by the orcas and their way of life. "Family is everything to them. You grow up in an orca family, that's it; you are there for life." Their diplomacy also amazes him. "They are the most powerful creatures in the ocean, yet they have evolved peaceful lifestyles.

"They are so successful. You know, we have been around for a few hundred thousand years as a species, and I don't know if we are going to make it much longer, given the way things are going. But they have been around actually in their modern form for millions of years.

"Take this little chunk of ocean right here. It kind of got going at the end of the last Ice Age, ten or twelve thousand years ago. When the rivers started forming and the salmon started showing up, that provided sustenance for them. And to think they have been here that whole period of time, and they are still successful. To me that is impressive, quite a bit more than how we are doing."

TOP: Orca families are societies of great antiquity, passing along language, food preferences, hunting methods, and other learning in a culture that endures from generation to generation. *(Dave Ellifrit/Center for Whale Research; taken under NMFS Permit 21238 and DFO SARA Permit 388)*

BOTTOM: The I16 matriline of the northern resident killer whale population. Orca whale families stick together in a vast ocean using calls that can be heard for more than a mile in quiet water. Orcas of the various types and populations in the Salish Sea do not interbreed or interact. *(Dr. Holly Fearnbach/SR3, SeaLife Response Rehab and Research; Dr. John Durban/NOAA; and Dr. Lance Barrett-Lennard/Ocean Wise Research, Ocean Wise Conservation Association; taken under Permit 2014-06 SARA-327.)*

Spong was among the first scientists who began to understand the complex emotional lives and intelligence of orcas. It was with the female captive orca Skana at the Vancouver Aquarium in 1967 that Spong, trained as a brain scientist, discovered he was interacting with a sophisticated mind. In experiments he devised to test her eyesight, Spong determined quickly that the orca actually was testing him. "Here I am a scientist, manipulating a subject, and I am being manipulated. That was interesting," Spong said. "What was she doing?

"I went through several phases to my experiment. In one of them, I set up an experiment that was entirely controlled from the other end of the building, and I didn't necessarily go to see her at all. I was just collecting data. She sort of ruined that for me by refusing point-blank to participate. That was the first time I ever thought about who she was. Eventually I thought, 'I better get to know her.'

"I would go down to the pool in the early morning and just hang out with her and sit on the little training platform on the edge of it and put my feet in the water and rub my feet over her head. I was figuring out how she perceives sound. There was a stainless-steel ladder that went down into the tank and I would ring it, and she would point her head right at the source of the sound the same way dolphins do. They hear through their lower jaw to their inner ear."

His interest in Skana grew by the day—along with their relationship. "I would go over there in the early morning and she'd come over, and at one point I decided to see whether I could walk out onto her back, and she'd let me do that," Spong said. "Then she would go out and bring me back to the platform. She would take me for little rides."

As his appreciation for the dimensionality of her intelligence grew, Spong no longer felt comfortable that she was in a tank. It wasn't long before he was speaking out against captivity, earning his employer's displeasure. He soon moved on, to fight orca captures and found OrcaLab.

By now Spong has logged thousands of hours of orca sounds and images and remains even more convinced that humans share space on this earth with beings the capacities of which we can only dimly begin to appreciate. "They gave me a life," he said of the orcas. "I feel very blessed and fortunate. The principal thing they have taught me is to love the planet. And they are so intelligent and generous. When you think about the issues we face, it is easy to abandon hope, but I don't think we should. The fact that they have survived a very long time—if they have, maybe we can too. We have similar age spans and families; there are a lot of parallels."

An Orca Sanctuary

One of the things Spong has learned in watching orcas' daily lives for so many years is that they are so much more than all business. The northern and southern residents alike are playful, athletic, and extremely tactile, continually touching and interacting in the water, with babies tossed by their parents and rolled over their backs. Orcas mate year-round, and their social lives are rich.

Drone photography has revealed whole new aspects of the lives of orcas, said Andrew Trites, the University of British Columbia researcher. He and his crew used a drone to film northern and southern resident orcas during research work in the summer of 2019. The footage startled the team when they crowded around a computer monitor for a first look. "It took our breath away," Trites said. All pretense to objectivity was lost in the wonder of what they were seeing.

Here were the orcas underwater, rolling and sculling along upside down just for fun, sliding along one another for the sheer pleasure of it. Particularly affecting for them all were the mothers and calves, always close together and so frequently touching. One mother nuzzled her baby, which slid all along her body for the joy of feeling her, then playfully tail-slapped her head. "Just like we would hug a friend or beloved, it's the all-important role of touch in maintaining bonds," Trites said. "Here we had killer whales reminding us of that. Whoever would have thought. I have been observing whales for so long, but I have never seen them like this."

Trites and the team spent two days observing the southern residents during that same research trip and was lucky enough to see J pod's newest baby. He watched amazed as the three-month-old, still nursing, nonetheless toted around a salmon for two days straight. Feeding entirely on her mother's milk, the fish wasn't food. Maybe she was teething or learning how to be a grown-up orca?

To Spong, the sensuality and the importance of pleasure and play in the lives of orcas had been most evident in watching the northern residents enjoy something believed to be unique to their culture. Northern residents gather at rubbing beaches along the forested slopes of Vancouver Island, using pebbles of a specific size piled in just the right depth for a full-body massage. The northern residents use these rubbing beaches—just certain ones, with exactly the right pebbles—as a spa, for belly rubs and back scratches.

Inside the lab, Spong showed me underwater footage from cameras at the Robson Bight (Michael Bigg) Ecological Reserve, set aside in 1982 just for the orcas. Here are some of the orcas' most-used, known rubbing beaches. Because of the wave action at one cove, the beach is piled deeply with the smooth round pebbles of just the size and texture the orcas apparently prefer. Here the northern residents eagerly take pleasures. I watched the video, enchanted as the orcas slid through pellucid aquamarine water, pressing the air from their lungs in silvery bubbles in order to sink low and scooch over the stones. Southern residents are not known to do anything like this. But these rubbing bouts—and these particular beaches—are clearly very important to the northern residents.

The reserve was created especially for these orcas and includes, in addition to the rubbing beaches, more than 3,000 acres of marine waters and 1,247 acres of upland forest, including rocky shoreline, headlands, and one significant bay, the Robson Bight. It is a place of abundance. The waters of the Tsitika estuary teem with all five species of Pacific salmon year-round. Clean, cold salt water offshore plunges rapidly to a depth of more than 1,300 feet into the glacially scoured center of Johnstone Strait. The estuary is home to abundant waterfowl and is a migratory funnel for salmon—and orcas feeding on salmon—heading to the Fraser River spawning grounds. Multiple shell middens also attest to the use of this special place by First Nations people for thousands of years, including one site big enough for several longhouses.

While there are other rubbing beaches in the orcas' territory, Robson Bight is a place the orcas are known to visit most from June to October, the

TOP: Southern resident orca mother J56 swims alongside her new calf. The baby has a salmon in her mouth, which she is just carrying around since she is still nursing. Perhaps this is some of her first training as a salmon hunter. *(Dr. Andrew Trites/University of British Columbia)*

BOTTOM: Northern residents enjoy a sanctuary created just for them in the Robson Bight (Michael Bigg) Ecological Reserve, where no boat traffic is allowed during most of the year. Along the shore, orcas rub on the smooth stones seemingly just for the delight of it. The reserve includes the beach, an important salmon stream, and the upland forest that protects it. *(Explore.org)*

main salmon migratory period for chinook traveling through Johnstone Strait. Here at the bight, their behavior is entirely different: there is less traveling, less feeding, and more resting, more play, more sex.

We went back outside, and Spong showed me around the grounds, hiking through the woods behind OrcaLab along a footpath subtle as an animal trail in the soft, mossy understory. Ravens' voices echoed in the quiet. Ferns reached shoulder high, and moss padded the rocks. Then all of a sudden we saw it, looming in the gloom: the Grandmother Cedar. At least one thousand years old, its vastness disappeared into the sky.

We walked around the tree, a journey that took many steps. Spong and I sat on a log, gazing at the tree, so large it seemed to make its own verdant world. The quiet was deep and soft as the duff underneath its branches, where tiny shoots and new life sprouted. Under the cedar's vast canopy, even the light seemed green. A small board was tucked inside a hollow at the tree's vast, spreading base, carved with the words Blessed Be He Who Leaves These Trees.

It was left from a decades-long fight, Spong said, as First Nations defenders of this land worked to

The Grandmother Cedar on Hanson Island is a just part of a landscape more like what the southern residents evolved with in Puget Sound. Big trees and intact forests nurture the salmon that feed orcas. The ecological health of a watershed is connected from the mountains to the sea. *(Steve Ringman/*The Seattle Times*)*

save this tree and this forest from logging. Today Hanson Island is an ecological preserve, managed by the Namgis people.

An Eagle's-Eye View

The next day, Spong motored us down the coast to OrcaLab's most remote outpost, at Cracroft Point. There, underwater cameras in fixed locations also discreetly record orcas in the wild. A shack atop a tiny platform on the point is the base camp for volunteers who track the comings and goings of the orcas and the human activity around them, especially in the Robson Bight sanctuary, just across the waters of Johnstone Strait.

At the shack, windmills and solar panels provide electrical power for the outpost. Volunteers Megan Hockin-Bennett and Shariana Manning had come halfway around the world from London to staff this little platform above the vast blue sea. It was their fourth year sharing this shack together for a stint with the orcas, sleeping on simple cots and cooking on a tiny stove in a galley no bigger than a card table. Hockin-Bennett ducked inside to make coffee and brought it outside to enjoy while watching for orcas.

We didn't have long to wait. Within moments, black dorsals slashed through the shine of sun on the water. The two volunteers abandoned their coffee and ran to spotting scopes to record which animals they were seeing, and their behavior, as best they could tell. The orcas zigged and zagged across the strait in a way they had not seen before. I marveled at the volunteers' commitment, camping on this bit of plywood out in the middle of nowhere to build, day by day, just a bit more of the record of the known lives of these orcas, in full knowledge that what they see from the surface, in the limited time they are with them, is just a glimpse of the orcas' wild lives.

The two were enthralled that summer watching a new baby with its mother among the northern residents. "To see a new generation, and everything she is teaching this baby; everything they do is based on teaching, from grandmother to mother," Manning said. "The matriarchs know where to go to get fish, where to rub on the beaches."

A collection of sea glass and the delicate dome of an urchin's shell, gathered over the years by volunteers, were placed on a sun-bleached plank by the viewing platform. An eagle cruised overhead, and low clouds spooled like skeins of wool through the almost black green of the forest across the water.

I went back inside the tiny cabin, enchanted as I looked around their world within a world, so snug and tidy with everything in its place, like the cabin of a boat. It was easy to imagine them tucked in here in their nest amid the blue waves and sky, day after day, keeping vigil with the whales.

A noise brought me back outside. With a rumble of their boat, wardens from the Robson Bight sanctuary were arriving to take us to the Eagle Eye Research Station, farther down the shore on a cliff across the strait from the reserve. From there, wardens Phil Rickus and Jessica Johnson kept watch over the reserve, to remind any boater approaching the rubbing beaches to back off, whether in a motorized craft or a kayak.

We headed in the wardens' sturdy inflatable Zodiac to the shore below the research station, the waves plumping the boat's soft bottom. At the shore I warily eyed the steep climb to the top of the cliff, with a rope ladder to begin the ascent. I was glad for

Orcas are incredible athletes, seeming to delight in hurling themselves in cartwheels over the water just for the heck of it. They live active, social, playful lives, constantly touching, horsing around, and exploring. *(Center for Whale Research; taken under NMFS Permit 21238 and DFO SARA Permit 388)*

it as we pulled our way up the trail, rocks skittering to the ground below us. The path was narrow, taking us clambering over outcroppings that opened to views of the water as we ascended: wider, bluer, higher, the sun flashing diamonds on the water's surface. Then, finally, there was the kingdom of the northern residents, resplendent below us.

The Eagle Eye Research Station is a rough observation station—just a long shed open on one side to the weather. Here, volunteers keep watch from this matchless vantage, looking for orcas. With the aid of spotting scopes, they note which orcas are present and monitor boat traffic. Using radios, they can alert the wardens down below to any unauthorized approaches into the orcas' sanctuary and rubbing beaches.

For the wardens, enforcement isn't just about keeping people out, but teaching why the sanctuary is important and what the orcas need to live their lives. "When they make it to the reserve, they finally have space; we see them change from traveling to socializing and beach rubbing," said Emma Weiss, a volunteer at Eagle Eye. "I feel this weight lifted, to see them in their reserve; people have to stop at the boundary line. It's important for them to have this break from all the traffic in the strait, to experience their full range of behaviors, not just dodging boats and hunting for fish. It doesn't look that big on a map. But it's incredibly important."

Conversation was cut short as orcas cruised into view, 130 feet below us. "It's a soupy mix of whales, really tight together. Look at all those dorsal fins," said Weiss, watching through binoculars. Even at this height, the water was so clear we could see the gleam of white pebbles shining through the depths. The sun was silver on the dark water and the forested cliffs velvet green, rising straight up to where we stood.

The sanctuary is not entirely protected or the surrounding area pristine. Fishing vessels are allowed in the reserve, a compromise struck by the British Columbia government. We saw seiners in the distance, working the sockeye run. Logging that is ongoing in the uplands adjacent to the reserve and its rubbing beaches has raised concern about the potential for siltation or landslides degrading this unique habitat so important to the orcas. And disturbance in Johnstone Strait can be dramatic, as we saw when suddenly, from around the curve of the point, came a cruise ship, five decks high, big as a building, coming down the strait from Alaska's Inside Passage, heading for Seattle.

The light was fading. It was time to go while we could still get down the cliff and the rope ladder to the beach, where we waited for Rob Williams to retrieve us at day's end and take us back to Alert Bay. It was sockeye season, and Ernest Alfred was eager to get back home, where his family was putting up cases of fish for winter and preparing for a potlatch just weeks away. The *Wishart* soon nosed near the beach, and Weiss fired up the Zodiac's outboard motor to ferry us over to meet it.

We told Williams about the seiners in the sanctuary and our surprise at the sight of the cruise ship in the strait, in acoustic continuity with the sanctuary even if distant from its boundaries. His reaction was matter of fact.

"It's not a pristine sanctuary, but I'll take what the whales can get," Williams said. He had worked his way through university as a backwoods guide and warden at Eagle Eye. He knew the sanctuary well, both its benefits and limitations. "You can't

Pacific white-sided dolphins seem to fly as they skim the surface northeast of Vancouver Island. Northern resident orcas live not just in another place but in another world, alive with wildlife. *(Steve Ringman/* The Seattle Times*)*

let the perfect be the enemy of the good, and the southern residents have nothing like this. It's past time they did."

He opened the throttle and we sped through seas of infinitely varied greens. As we passed a small island, I saw a bear on the beach turning over rocks, looking for crabs. "The farther north you go, the wilder it gets, and the less people; we think that is why the northern residents are doing well," Williams said.

What Works There Is Needed Here Too

I thought about Tahlequah carrying her baby past the coal docks south of Vancouver piled high with coal for export, ghosting black dust into the water. She had journeyed through the shipping lanes in Haro Strait and amid the container ships and oil tankers towering over her. Even in her primary foraging grounds near San Juan Island, she swam through busy boat traffic, including commercial whale-watch boats out to entertain tourists from morning until night.

The southern residents also contend with more polluted water—toxics that seep into the food chain and then into orca mothers' milk. A September 2018 paper found that PCB effects on reproduction and immune function threaten the long-term viability of more than 50 percent of the world's orca populations, especially those nearest urban areas.

A group of toxic man-made chemicals, PCBs were banned from manufacture in the US in 1979 but are still present everywhere in the environment. PCBs were used around the world for decades, primarily to insulate and cool electrical equipment and prevent electrical fires. They were also used in hydraulic systems, lighting and cable insulation, paint,

OPPOSITE: Orca A60 breaches in Blackfish Sound, a vital feeding area for northern resident orcas seeking salmon in the inside waters off northern Vancouver Island. *(Dr. Andrew Trites/University of British Columbia)*

RIGHT: A northern resident killer whale swims with a pair of white-sided dolphins. Resident orca diplomacy is so sophisticated the orcas are able to share space with many species of marine mammals in the Salish Sea and outer coast without conflict. *(Dr. Holly Fearnbach/SR3, SeaLife Response Rehab and Research; Dr. John Durban/NOAA; and Dr. Lance Barrett-Lennard/ Ocean Wise Research, Ocean Wise Conservation Association; taken under Permit 2014-06 SARA-327.)*

caulking, sealants, inks, and lubricants. Today PCBs still leach, leak, dissipate into the atmosphere, and contaminate runoff during rainfall.

Southern residents are at graver risk of health effects from PCBs and other toxics than the northern orcas, the study found, because the southern residents eat more contaminated prey. Toxics also are more dangerous for them because they don't always have enough to eat. When they go hungry, orcas burn their fat, releasing more toxics into their bloodstream. To the north, life is better for the orcas—and babies tell the story. A total of ten calves were born to the northern residents in 2017, and eight more were born in 2018. Tahlequah's baby, born in August 2018, was the first for the southern residents in three years, and she lived for only half an hour.

I went out with Rob Williams and Erin Ashe in the *Wishart* one last day, hoping to spend it with the Pacific white-sided dolphins Ashe was studying, but a 15-knot wind was blowing. Ashe remained optimistic, searching the sheltered coves she knew well. But we were skunked, cove after cove, as the boat slammed up and down in the waves. "Just a few more bumpety bump bumps," Rob said cheerily to daughter Clara, who was taking it all in stride as I clung to my seat with both hands; even Molly the Chesapeake looked a bit woozy.

Then my trip with one intrepid family encountered another as a group of orcas cruised into view, including a baby that couldn't have been more than a few months old, working hard to keep up. Its head high in the water, the baby fought the wind and waves, staying close by its mother in her slipstream. In no time, the orcas surged past us, flinging their bodies easily through the spray and rollers, masters

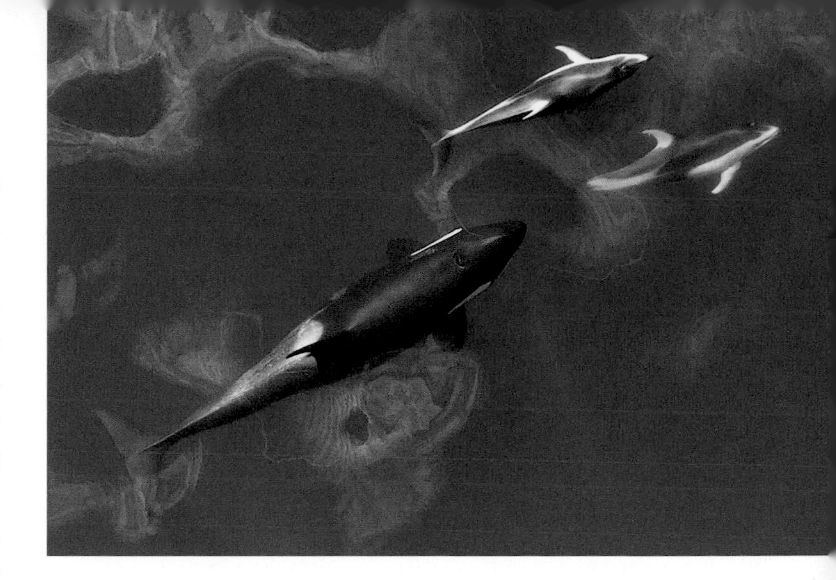

of their realm. Williams was right, and the northern residents' steady population growth proved it. If they have what they need—even if it isn't perfect—these animals do just fine.

That was both a heartening and a sobering realization. For orcas and the salmon they depend on to win in the more populated, developed home waters of the southern residents, we have to *let* them win. That means stepping back and making space for species other than our own.

Environmental restoration is a practical matter of reviving the physical and biological processes animals need to survive. It takes setting the right priorities, paying for habitat that needs to be fixed, and not wrecking more. This is something we know how to do, and are doing, all over the region. From birds to bugs to forage fish to salmon, species are rebounding where they have what they need, even when conditions that allow them to thrive were greatly depleted for more than a century. Restoration won't be possible everywhere, of course. But the good news is that already we have plenty of evidence from our own experience that when nature is given a chance, life surges back.

It's a matter of doing the work.

CHAPTER SEVEN

The Work

The world's largest-ever dam removal project, on northwestern Washington State's mighty Elwha River in the Olympic Mountains, has proven even to the skeptics that change is possible and can happen more quickly than anyone imagined.

The first place I ever camped in the Northwest was along a tributary of the Elwha in Olympic National Park, and its song at Camp Lillian almost made me

During the last summer before dam removal began in September 2011, these Elwha chinook returned for their fall run to their hereditary spawning grounds, only to be blocked by the Elwha Dam just 5 miles from the river mouth. Even after a century of futility the fish never stopped trying to come home. *(Steve Ringman/*The Seattle Times)

LEFT: The late Adeline Smith, age ninety-three in this photo, was one of the oldest members of the Lower Elwha Klallam Tribe. She said she had seen a lot of changes in her life. But the dams coming out? She said, "I never thought I would see the day." But she did, just before she passed in 2013 at the age of ninety-five. *(Steve Ringman/*The Seattle Times*)*

OPPOSITE: Pink salmon rest in a small pool on their way up the Dungeness River to spawn. Greatly depleted due to the dams, pinks were once the most numerous run in the Elwha River, with more than one hundred thousand fish bringing a crush of biomass and nutrients back to the watershed from the sea. With both dams gone, river scientists hope pinks will take the river by storm again one day. *(Steve Ringman/*The Seattle Times*)*

forget how cold I was in my summer sleeping bag, which was not the right choice for an autumn overnight in the backcountry. But no matter—the Elwha is worth shivering for.

I've returned over and over to this river since the early 1990s, been all over its two former dams and their powerhouses, and met some of the people who ran them. I met even more people who longed for the day these dams would be taken out, none more so than the Lower Elwha Klallam Tribe, who lost the most and gained the least from the building of two hydroelectric dams on the Elwha.

The Lower Elwha Klallam people's aboriginal territory stretches over a sweep of the Olympic Peninsula and across the water to Canada. The tribe initiated the push to remove two dams on this river: Elwha Dam, built 108 feet tall in 1910 with no fish passage, just 5 miles from the river's mouth, and Glines Canyon Dam, built 210 feet tall in 1927, 8.6 miles farther upstream.

When the Crown Zellerbach Corporation, owner of the Glines Canyon Dam, applied to the Federal Energy Regulatory Commission in 1973 to renew the license for the dam, the tribe and its advocates saw opportunity. Their work would culminate in 1992, with passage of an act of Congress calling for dam removal to restore the river's legendary salmon fishery. It would take decades and many political

twists before removal finally began. I well remember that day. I had hiked into the woods to hear the first bang of a hydraulic hammer on the concrete of Glines Canyon Dam. Before long, I heard an unmistakable racket: Demolition was underway. The date was September 15, 2011.

An official ceremony marking the event was held two days later, and some of the Elwha tribal elders who had worked the longest to watch this day come, including the late Adeline Smith, finally got to see it. The fish showed up, too, circling at the foot of the dam that was still in their way—but not for much longer. "I never thought I would see the day," Smith told me. "There," she said, pointing at the snowy Olympics above the dams. "They are going to go up there."

After all the years of talking about it and fighting over it, the takedown of Elwha Dam happened fast. With a combination of blasting and teardown with heavy equipment, by five in the evening, March 9, 2012, Elwha Dam was gone. Heavy equipment operators say they were shocked to see salmon pressing up into the river above the former dam site right away, despite water thick with sediment.

These fish were a collective triumph of a grateful region that solved a complex economic and ecological problem through a long public process. The work ultimately yielded not only salmon but proof that we can work together to make things better. And that given the chance, nature will respond.

The Grand Experiment

The Elwha Dam removal and river restoration project is a $325 million grand experiment, a rare opportunity for a watershed-scale renewal, from the mountains to the sea. More than 83 percent of the 321-square-mile watershed is still pristine, and permanently protected, in Olympic National Park.

LEFT: After dam removal on the Elwha a great pulse of sediment trapped behind the dams made its way downriver all the way to the sea, where it has created Washington's newest beach. Natural movement of sediment is crucial to the health of a river and its estuary and near shore. *(Steve Ringman/*The Seattle Times*)*

OPPOSITE: Young herring by the thousands school amid lush kelp forests off the beach near Freshwater Bay, west of the mouth of the Elwha River. Forage fish need natural beaches for spawning; they feed the salmon that feed the orcas in an interconnected circle of life. *(Steve Ringman/*The Seattle Times*)*

Tearing down the dams reopened 70 miles of spawning habitat to all five species of Pacific salmon. It also restarted the 45-mile flow of sediment and big wood in the river all the way to the salt water. The Elwha is a migratory pathway not only for fish but for large woody debris, sand, and gravel from the high country that ceaselessly seek a hungry sea. Beaches and the nearshore must be fed this material or they starve to bare cobble. The Elwha restoration is reconnecting these vital cycles.

An interdisciplinary team of scientists from the Lower Elwha Klallam Tribe, academia, nonprofits, and federal and state agencies has been recording the condition of the river before and after dam removal. They have found that the river quickly adjusted to the sediment released, with its water quality returning to baseline clarity sooner than anticipated. The river already more closely resembles the other wild rivers of the Olympic Peninsula that were never dammed, such as the mighty Queets and Hoh. The bare cobble of the beach and near shore at the mouth of the Elwha today are covered in soft sand.

No wonder: Dam removal increased by two orders of magnitude the amount of sediment in the Elwha during the first five years of recovery, from 2011 to 2016, Jonathan Warrick of the US Geological Survey and his coauthors reported in a

paper published in 2019 in the journal *Nature*. More than 741 million cubic feet of sediment—enough to fill nearly two million dump trucks with gravel, cobble, silt, and clay—hurtled down the river, with 90 percent of it discharged to the coast. Most of it was transported offshore of the river delta by the prevailing current and dispersed. But plenty ended up on the beach too. Here was a shoreline that had lost more than 520 feet of beachfront to erosion between 1939 and 2006. With dam removal, a cobble terrace in the nearshore waters was transformed to Washington's newest beach.

As Ian Miller, coastal hazard specialist for Washington SeaGrant, astutely observed in his quirky, wonderful blog, the *Coast Nerd Gazette*, today the beach even *sounds* different. Close your eyes, he exhorts, and hear the change, from the clacking of bare rock on rock to the sound of waves on a sandy beach.

Matt Beirne, natural resources director for the Lower Elwha Klallam Tribe, watched it happen, in dive surveys undertaken with several agency partners to track the changes in the near shore. "After the dams were removed, you had this wave of sediment released into the Strait of Juan de Fuca," Beirne said. "There were times we would be down at the bottom and the sediment plume would shift and completely envelop you, and things would just go completely black. You just need to be calm."

Survey markers sunk on the bottom 37 feet down in 2012 were covered by only 7 feet of water a year later, because 30 feet of sand had piled up. "It was this pure, fine, beautiful sand, undulating downward toward the strait," Beirne said. The change to a sandy bottom from bare cobble set off changes in the community of animals using the near shore. "The thing we noticed first was, the sand lance were everywhere," said Beirne.

Sand lance are key forage fish that convert the plankton of the sea into the meat and fat that seabirds and fish—including chinook salmon—feast on. Sand lance live in the soft, sandy bottom of the nearshore environment and need sandy beaches to spawn. As the divers skimmed along the bottom in their survey, they saw sand lance everywhere, poking their heads out of the new drifts of soft sand. "You couldn't count them all," Beirne said. "It was just bristling with life."

That was just the beginning. On that same dive, the team saw an octopus cruising along; then the next year, the first wave of juvenile Dungeness crab, no bigger than a dime, were moving in. In 2014, a mass spawn event at the mouth of the Elwha of eulachon, another important forage fish, brought in predators by the score: marine birds, dozens of seals and sea lions, and at least fifty bald eagles. "I can remember once seeing fifteen. But fifty—there was just this feeding frenzy; it was phenomenal," said Beirne.

Forage fish—including eulachon, sand lance, surf smelt, and herring—are nutrition powerhouses for the salmon that feed orcas. "No little fish, no big fish, no blackfish" is a Coast Salish saying for a reason. Without stoking the food web that nourishes salmon, orcas will not survive. Salmon also are coming back to the main stem and tributaries upstream of where the Elwha dams once stood. Steelhead, chinook, and coho are back in reaches of the river they were shut out of for more than a century.

A Cautionary Tale

While dam removal has boosted the flow of sediment and sand to the near shore, the Elwha is both a success story and a cautionary tale. A more than 2-mile-long bulkhead runs along the foot of the bluffs back from the beach on the east side of the river mouth. The bulkhead, built in the 1950s, armors an industrial water pipe and was extended again in 2007 to protect a city landfill. But this barrier also blocks the natural flow to the beach of gravel, sand, and sediment embedded in those bluffs since the glaciers dropped it there, as surely as the dams blocked the flow of fish up and down the river.

This is the kind of ordinary shoreline infrastructure that people have built for decades all around Puget Sound, including concrete walls, boat ramps, riprap, pilings, and shoring. It can be quite harmful to the flow of sediment to the beach and near shore and to the lives that habitat supports, such as surf smelt, another crucial forage fish, that spawns only on sandy beaches. Starve the beach and near shore of sediment and sand, and it starves the food chain, all the way up to the southern resident orcas.

The disturbance goes beyond the immediately hardened area, University of Washington biologist Megan Dethier and her coauthors found in a 2016 paper assessing the cumulative effects of bulkheads in the Salish Sea: everywhere the beach would have been fed by an unarmored bluff, it is starved.

An estimated 223 miles of Puget Sound's original 657 miles of so-called feeder bluffs have been

CLOCKWISE FROM TOP LEFT: Bull kelp forms forests off the coast of Washington, providing important nursery grounds for young fish. *(The Seattle Times)*; Herring spawn colors the water in Hood Canal in a display of otherworldly beauty. The herring spawn is a rite of spring that occurs in nearshore waters from south Puget Sound to British Columbia, feeding a conglomeration of seabirds, marine mammals, and fish that throng for the feast. *(Steve Ringman/*The Seattle Times*)*; Like tiny jewels, herring eggs stick to eelgrass and other aquatic vegetation. *(Steve Ringman/*The Seattle Times*)*; A bald eagle sits above the new layer of sand at the mouth of the Elwha River. The Lower Elwha Klallam Tribe is fishing crabs off the mouth of the Elwha River for the first time in generations. *(Steve Ringman/*The Seattle Times*)*

cut off by riprap and other barriers. The relative length of armored feeder bluff is much more extensive in the more developed areas of King, Pierce, and Kitsap Counties and less prevalent in northern and western areas such as Whatcom and Clallam Counties. But in the Seattle central waterfront, more than 80 percent of the coastline is armored.

Every beach counts, because for the southern residents, every salmon counts, and salmon have to eat too. "The concept of saving the orcas without saving the food web, I'm sorry, that is not going to work," Dethier said. It's all connected: Orcas need salmon, which need forage fish, which need natural beaches to spawn.

To get a sense of how an undisturbed, natural nearshore habitat looks—and functions—I went snorkeling just west of the mouth of the Elwha with Anne Shaffer, executive director of the Coastal Watershed Institute.

Waves rocked and swelled the Strait of Juan de Fuca as I snorkeled at the surface. I tipped my face up to look for the seabirds that would guide me as, like them, I searched for herring, another key forage fish. The gulls wheeled and spoke their wild, pelagic cries.

Just offshore, I floated through a kelp bed, thick and lush, its stipes, or stems, bearing shiny bronze fronds growing from the sea bottom toward the sun. Kelp, a large brown seaweed, attaches to bedrock or cobbles at the sea bottom. In dense stands such as the one I was in, it forms a true underwater forest, with an understory as well as a canopy. The kelp bed softened the swells as I lolled in the blades and stipes all around me, slipping through it like a seal. Scanning the depths, I saw a shine in the distance and swam to it; this is what I had come here hoping to see: a vast school of young herring.

The fish were uncountable, a multitude swimming as one organism. I floated above them silently, barely moving, just watching the thousands of silvery lives and ever-changing colors. The sun spangled on diatoms adrift in the water and refracted from the moving silver torrent of fish. A starry flounder swam through the school, heading on some urgent flounder errand.

With twenty-three species of kelp, Puget Sound is home to one of the most diverse kelp floras in the world. Kelp creates a nursery where young fish hide and flourish, and herring spawn. Hundreds of other species also live in the holdfasts of kelp and along their stipes. Some fish use the kelp forest for just part of their life, while others remain there for a lifetime.

Kelp is remarkably tough; the forces that kelp forests experience in currents and wave surge, as they are pulled and released, are comparable to a terrestrial plant subjected to wind velocities of well over 400 to 500 miles per hour, said David Duggins, supervisor of marine operations and senior research scientist at the University of Washington's Friday Harbor Labs on San Juan Island.

Kelp also is a fundamental food source, providing nutrition not only to animals that graze on its surface but also to the water, as the kelp shreds and decomposes, Duggins reported in *Science* in 1989. Detritus from kelp feeds everything from sea anemones to shrimps and crabs, starfish and cormorants.

The herring I swam with were the young of the year. They had grown from eggs deposited that spring on all kinds of surfaces, from kelp to emerald-green eelgrass beds that root in the soft, sandy bottom of sunny shallows closer to shore, in quieter water.

There are about 57,000 acres of eelgrass in Puget Sound, with about half growing in small meadows at the edge of the shoreline and the remainder in broad tidal flats. Across Puget Sound, the total area of eelgrass beds has remained relatively stable since 2000. But that overall picture conceals local losses. Sometimes docks that shade eelgrass are to blame, or destruction is due to dredging, anchoring, or other physical damage. Warming waters also are fostering a wasting disease that is killing eelgrass.

Quilcene Bay, along Hood Canal, is one of the herring strongholds in Puget Sound. In the spring of 2018 I got word of a big spawn underway there and headed over for a look. I could hear the abundance of animals gathered for the feast before I saw it: barking sea lions, the clattering call of kingfishers, the chitter of an eagle in the forested banks of the cove. Cormorants and seagulls patrolled for herring, and sea ducks massed, peeping and calling. An eagle swooped from a fir tree and hit the water, then rose with a silver herring between its talons. It flew back up to the treetop to shred its meal.

The beach was a soft curve of sand, backed by a forested natural bluff. In the shallows, eelgrass rooted, emerald and thick. It slowed the current and dampened the waves, trapping sediments and detritus, nurturing a marine menagerie of tiny lives, the copepods and amphipods—crustaceans and aquatic bugs that feed young fish. Some of the blades were so overgrown with life that they were fuzzy.

Along the tide line on the sand, a tangle of eelgrass was heaped by the waves. Bending to look more closely, I saw every blade of grass was covered with tiny herring eggs. Each was translucent and pearly, no bigger than the tip of a crayon. I held a piece of the seaweed to the sunlight to appreciate this tiny beginning of everything: of herring, of salmon, in a direct line of nutrition all the way to orcas.

Herring eggs are packed with fats and protein, perfect capsules of energy and nutrients. Pacific herring are a vital food source for everything from humpback whales to porpoises, salmon, sea lions, seals, surf scoters, scaups, wigeons, harlequin ducks, and rhinoceros auklets. Scoters will strip herring eggs off Sargassum weed like corn on the cob. So important is the herring spawn as a food source that some animals, including horned grebes and surf scoters, time part of their reproductive cycle to be in sync with the herring spawn.

The spawning season for herring is a Puget Sound rite of spring. It begins as early as March and continues in a six-week progression throughout the Salish Sea. These spawning events link the coastlines, from Olympia to north Puget Sound, to the Strait of Georgia in British Columbia. The nearshore waters are tinted in tropical colors of teal, pale green, and aquamarine as the spawn swirls, colorful in the water as northern lights in the night sky.

The release of billions of eggs and great clouds of milt comes just as the water shines green with chlorophyll when the spring sun detonates a plankton boom. This is the pasture that will feed the hatching baby herring and sand lance and other forage fish.

These baby herring will be teeming in the nearshore waters just as young salmon, transitioning to a diet of adult food as they leave the Elwha, will need them to fatten up for their journey out to sea.

TOP: An American dipper flies above a side channel of the Elwha River. Dam removal is benefitting not only salmon but also birds and other wildlife that feast on salmon and salmon eggs. Since the dams came down, dippers in this area are bigger in body size and are bearing more young. *(Ken Lambert/*The Seattle Times*)*

BOTTOM: After the world's largest-ever dam removal project on the Elwha River, salmon, including these chinook, have returned to the river in greater numbers, feeding birds, bears, and even orcas hunting where the river meets the sea. *(Steve Ringman/*The Seattle Times*)*

Elwha Resurgent

Salmon restoration was the big headline in the Elwha dam removal story, and the southern resident orcas are its celebrity beneficiary. But many other less celebrated animals are affected by dam removal and are themselves driving the results.

A host of animals are colonizing habitat, pollinating plants, dispersing seeds, aerating soil, breaking down and redistributing nutrients, and building soil, according to a 2018 paper published in the journal *Ecological Restoration*. Animals are connecting the established forest to the river and aiding in replanting some 700 acres of former lake beds. Their scat and seed caching are helping to boost the conversion of the lake beds' bare, blowing fine silts to a landscape alive with trees, shrubs, flowers, birds, and bugs. Mice were among the first brave pioneers of the lake-bed silts, lead author Rebecca McCaffery, of the USGS Forest and Rangeland Ecosystem Science Center's Olympic Field Station in Port Angeles, and her coauthors found in their research.

Weasels, voles, shrews, wood rats, and chipmunks also were among the earliest to utilize the former lake beds during the first two years after dam removal, the authors learned. Signs of beaver activity were found on both reservoir beds, most often in cuttings of cottonwood and willow and in dam-building activity on small tributary streams. Black bears, black-tailed deer, and Roosevelt elk all have been seen using different areas of the reservoir beds.

All of these animals helped give a boost to the extensive replanting efforts on the lake beds. Unprecedented in its scale, the replanting, which concluded in 2017, already has resulted in a young forest grown more than head high and vegetation so thick in places it is impossible to see the river. Swaths of purple river lupine hum with bees, and alder and willow are loud with singing birds.

The young trees meet the original old-growth forests beyond the banks of the former reservoirs in a blur of green, creating a seamless cover of the once-bare silts of the lake beds. It won't be long before visitors won't even be able to tell they are walking where the reservoirs would have once been over their heads. Birds have colonized the rising young forest and shrub communities. River otters are expanding their range into the middle river. Elk are sauntering through winter range returned to them along the banks.

One of the most important but unseen changes since dam removal is the return of nutrients to the Elwha that has come along with the increase in the number and range of salmon in the river. That is also boosting juvenile salmon growth and benefitting other animals, Chris Tonra, professor of avian wildlife ecology at Ohio State University, and his coauthors found in a 2015 paper in *Biological Conservation*.

Tonra studied American dippers, an aquatic songbird, briefly trapping them in very fine nets strung across the Elwha, to sample their blood, claws, and feathers before and after dam removal. Dippers not only eat insects that prey on salmon carcasses but also eat salmon eggs. After dam removal, Tonra documented increased nutrition levels in the birds, derived from salmon. The dippers with access to salmon were bigger in body size and twenty times more likely to raise more than one brood of young. Their fledglings were in better condition, and the survival rate of adults was improved.

These results mean that marine-derived nutrients had already made their return to the ecosystem, traced directly to the salmon that came back with dam removal. And that shows that both aquatic and terrestrial food webs have the capacity to be restored quite rapidly, despite the river being dammed for a hundred years.

Return of the Kings

While Elwha recovery is not only about salmon, they sure were exciting to watch come back up the river during the chinook migration of 2018.

I followed Mel Elofson on a footpath through cedars, firs, hemlocks, and deep thickets of salmonberry and sword fern. It felt good to be here, in this classic lowland temperate rain forest, growing lush and thick, overhanging the banks of the Elwha. Elofson is assistant habitat manager at the fisheries department for the Lower Elwha Klallam Tribe. We had come to the river to marvel at Elwha chinook, some of the biggest in Puget Sound: still here, just like his tribe.

It wasn't long before we saw what we'd come looking for that day. It was late in the year, and the river was so low that the fins of the mighty chinook were sticking out of the river as they swam up the current. We had come just to watch—and listen.

The glimmer of spider silk shone in the slanting autumn sun, and a kingfisher clattered. Clouds of insects hummed. As I looked at the bank near my feet, I saw in the silty bottom the skull of a spawned-out salmon, its toothy jaws agape, showing rows of snaggle teeth. The sounds of life were everywhere: the splashing of salmon thrashing up the river, the feasting birds and bugs. Mel leaned down and picked up an eagle feather, tucked among the moss and leaves. He saved it as a memento. He said he had recently seen a bear from his house, stalking the river for fish—the first he had seen in a long while.

The population of chinook prior to dam removal averaged 2,900 returning adults each year. Since dam removal was completed on the lower dam in 2012, the number of returning adults has ranged from an estimated low of 2,628 in 2016 to an estimated 7,600 in 2019. Coho smolts have boomed, and steelhead are making a comeback—even wild summer steelhead have been seen above the site of the former Glines Canyon Dam.

This is a true Lazarus story. Summer steelhead on the Elwha had become so rare they were all but extinct. But with the dams out, the tens of thousands of rainbow trout that had been locked up above the dams had their chance once more to go to sea. Their anadromy was reawakened, resulting in a silvery cascade of oceangoing summer steelhead. Just three years after dam removal, the Elwha River population of summer steelhead was bigger than that of any of Washington's coastal rivers, and already in 2019 numbers were estimated at about 920 fish.

John McMillan, science director for Trout Unlimited's wild steelhead initiative, can't get over the near miracle of these fish. He has regularly snorkeled the Quileute and Hoh Rivers and said he's never seen more than 135 steelhead in either watershed in a quarter century. Puget Sound rivers also have no populations that even come close. Columbia River steelhead are tanking. And yet, even during some of the worst ocean survival conditions since the 1990s, summer steelhead in the Elwha are going gangbusters. "It's the most exciting discovery

TOP: The wonder of orcas never grows old. Bubbles briefly cover this southern resident's pectoral fin as it surfaces in Commencement Bay near Tacoma in November 2018. *(Steve Ringman/*The Seattle Times*; taken under NOAA Permit 21348)*

BOTTOM: Orcas often hunt for salmon in kelp beds and seem to enjoy playing in the long stipes and slippery fronds too. *(Dr. Astrid van Ginneken/ Center for Whale Research; taken under NMFS Permit 21238 and DFO SARA Permit 388)*

I have had in science," McMillan said. "It's a shock these fish came back."

Bull trout are growing bigger, faster, too, and accessing more of the river. Lampreys are back in the river.

Conditions for fish are expected to just get better. River sediment loads have returned to normal, and that helps not only salmon but macroinvertebrates that feed them. Dam removal detonated a near-total wipeout of mayflies, stone flies, and caddis flies in the river, all prime salmon food. Sarah Morley of NOAA's Northwest Fisheries Science Center, who sampled fish diets before and after dam removal, learned that fish switched diets when sediment loads were highest, to eat terrestrial bugs—and even other fish.

There is still a long way to go before recovery goals for the river are met. The chinook in the river are still nearly all hatchery fish, and only small numbers of adult chinook are making it into the upper watershed. Scientists are trying to figure out why—and what to do about it, if anything. Blasting rockfall out of the river hasn't solved the problem. Perhaps it's just a matter of time. Recovery throughout the river is still at its beginning stages.

Fishing is still a long way off—at least for people. But not, apparently, for the orcas.

Once back from my trip with Elofson, I heard from Ken Balcomb of the Center for Whale Research. He and other scientists survey the orcas repeatedly every year to determine their population and health. Balcomb keeps a second home in Port Angeles, and when he got word of an orca sighting, he took off to go see what he might find. In my phone were the photos he sent me of the orcas he saw that day.

Southern resident orcas, feeding off the mouth of the Elwha.

The High-Class People

Elwha restoration is the work of many hands, and in that there is a lesson. In the work ahead for orcas, restoration can't be imposed, but it can be embraced by the broader community where it makes sense—even when the changes involved are very large. It took a lot to get there, but the Elwha restoration shows what is possible.

This is everyone's work, and what we each bring will depend on what we have to offer. It is the offering that is the point. I learned that from Raynell Morris of the Lummi Nation.

In the summer of 2019, tribal members were dismayed by a summer with low chinook returns, and they took to the water five times that season to make ceremonial offerings of chinook to the southern resident orcas. I went along on one of those trips.

As we set out at sunrise, Morris, then senior policy advisor in the Lummi's treaty protection office, called on her cell phone to the skippers she knew were already out on the water. "Do you have any fish?" she asked, time and again. Skipper after skipper reported that they had no salmon to offer, no catch at all. As we continued on, we saw more oil tankers than fishing boats in the Salish Sea. With catches so low, fishermen couldn't justify the cost of going out.

Richard Solomon, a Lummi spiritualist and lifelong fisherman, prayed as skipper Aaron Hillaire of the Lummi police department, piloting the tribal police boat, scanned the water with binoculars. But he did not sight what even prayers could not raise. As we traveled the southern residents' traditional

The southern resident orcas have defied shooting, captures, pollution, development, salmon declines, noise, and disturbance to persist in a changing world. *(Mark Malleson/Center for Whale Research; taken under NMFS Permit 21238 and DFO SARA Permit 388)*

summer waters, we did not see a single orca, hour after hour.

It was an unprecedented summer. The southern residents had barely been seen yet that year in the San Juan Islands, when typically in the past they would have been out foraging for fish amid the islands, particularly on the west side of San Juan Island, every day. But with few chinook to chase, the orcas apparently could not justify the energy of coming here either.

"They are looking, just like we are looking," Solomon said of the orcas and Lummi fishing families. "Where are they supposed to go? How are they going to feed their young ones when they can't even find fish? And how are we going to show our children how to live when we have no fish? We are the same."

Eventually we landed at a traditional village site of the Lummi people on Orcas Island, where a layer of shell midden—the remains of thousands of years of cooking and eating—was visible, in bits of broken white shell some 10 feet thick along the shore. We docked and headed to the midden beach. This would be where Solomon and Morris would make their offering to the orcas.

I watched as they walked single file to the beach, noticing that Solomon was gathering blackberries but not eating them. He had put on a cedar chief's hat, made by his sister and adorned with an ermine tail. On either side of his head, reaching to his collarbone were the long white tassels of the wool headband he also wore, made by the late Lummi hereditary chief Tsi'li'xw Bill James. In one hand, Morris carried green cedar boughs she had brought

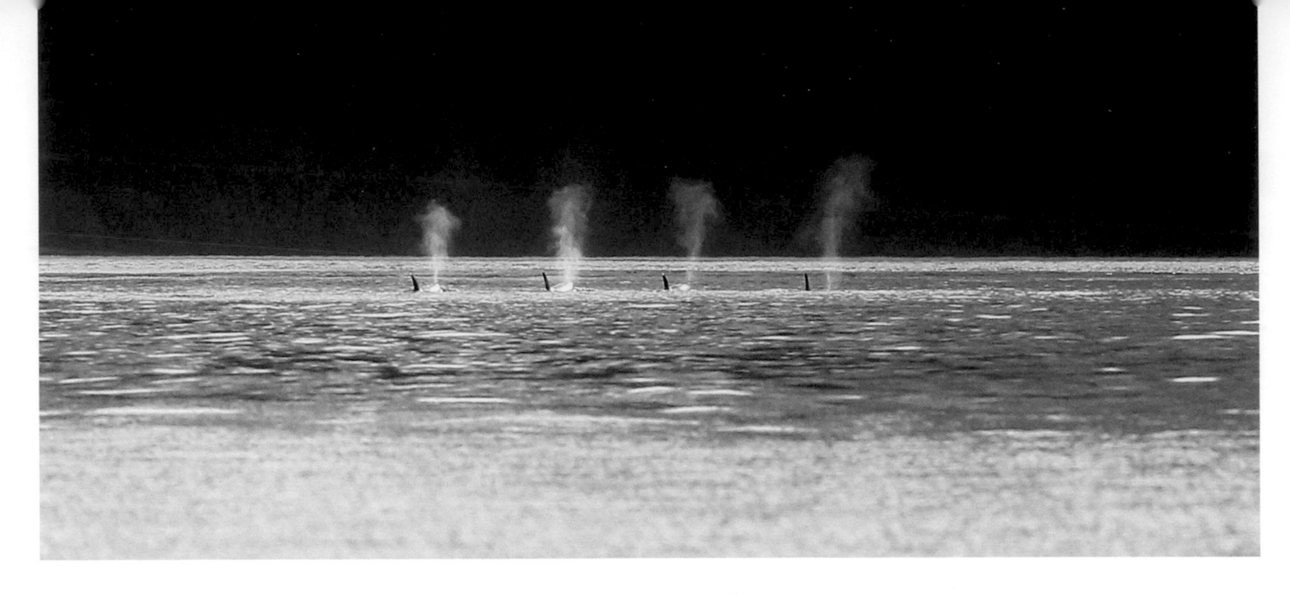

LEFT: Orcas have been the top predator in every ocean of the world for more than six million years. They can survive and even thrive if they have what they need. *(Mark Malleson/Center for Whale Research; taken under NMFS Permit 21238 and DFO SARA Permit 388)*

OPPOSITE: Mother orca Tahlequah nuzzles her baby, J57, born in September 2020, a spark of new hope for the southern resident orca families. *(Dr. Holly Fearnbach/SR3, SeaLife Response Rehab and Research; and Dr. John Durban/SEA, Southall Environmental Associates; taken under NMFS Research Permit 19091)*

from the reservation, and in the other, a ceremonial deer-hoof rattle. Once at the beach, they walked to the water's edge and made their offering: blackberries floated on cedar boughs, sent with their prayers and a song, punctuated with the beat of the rattle.

As we walked quietly back to the boat, Solomon paused to sit on a low wall by the water and face the cove. He raised an honoring song passed down from his grandmother, about the high-class people, as she called them. Orcas.

It was a song for the orcas that had passed away, the *qwel lhol mech ten* that did not make it: "The high-class people are going to sleep now," Solomon sang, over and over, first in his language and then in English, raising his arms in praise, his voice at the edge of a cry. "I just hope they heard me," Solomon said, then paused, as if listening for an answer from the other side. "I am sure they have," he said.

We continued walking quietly to the boat. "I hope they come back and tell us what to do," Solomon said. "That's all we can do."

The prayers they had offered that day were for J17—Tahlequah's mother—and K25, both southern resident orcas known to be dangerously thin. "We are not ready to call it a memorial yet," Morris said of the ceremony on the beach. But by the end of the summer, both J17 and K25 and a third southern resident orca from L pod all were dead.

As we started our trip back, Morris spoke of the importance of doing what they could for the orcas that day. The emphasis was not on what they could not do or did not have. It was on the effort they had made and offering whatever they could find to give—if not salmon, then berries, prayer, and song. In these home waters without orcas or salmon, it seemed the right message.

We approached a fishing site near the tribe's reservation, where the clear, clean, cold water was deep green and still. Too quiet. No one was fishing; the season was curtailed. Hillaire switched off the boat's engine so everyone could just sit and be still for a while before returning to shore. What a long, sad summer, the second in a row, with a total of six more southern residents dead.

"It is incredible," Morris said of the empty reef-net platforms, the dead orcas. "But it is not surprising. We share this pain with the *qwel lhol mech ten*. We share this place with the *qwel lhol mech ten*, with the salmon. The ownership of what happened to this land and this water, everyone needs to own it.

"We've done this. And it's our job to heal it. We allowed this. It's not their work or your work. It's all of our work."

On the final miles of the ride back, everyone was quiet; there was a sense of peace. For on that day, in their way, each person had done what they could and what they were called to. Morris said she felt hopeful, that this is what we all are here to do: let the orcas lead us home, to rebuild the enchantment and abundance that is the natural condition of this place. "Famine. There is no Lummi word for that," she said.

There is a consilience of her traditional knowledge with science. The home of the southern resident orcas and the salmon is our home too. And when they thrive, so will we. Perhaps, then, there is a very simple test to determine—if we do the work of restoring and protecting the waters and lands needed by orcas and salmon—whether we are getting somewhere. It's a metric that won't appear in any scientific paper. But it will make sense to people who read all those stories and cared about Tahlequah in 2018.

They saw that picture of the mother orca with her dead baby calf on her head, and many responded to her loss with primal grief, not only because of what it meant for her but because of what it meant for us. It took a mother orca named Tahlequah to raise up the plight of orcas and salmon and the imperiled abundance of our common home in a way that is beyond politics or even words.

Tahlequah's healthy calf is a new chance for her, and for us. Her persistence offers a simple but powerful five-word challenge, an anthem and a work song for a region with much to do, not only for her and her family but for the future of a Pacific Northwest alive with orcas and salmon:

May this next calf *live*.

ACKNOWLEDGMENTS

After a book like this, built over many miles and years, I have many people to thank. I'll start with the scientists who took me out on their research trips where I saw the orcas, heard them breathe, and was confronted with their eminence.

So thank you to Ken Balcomb and Dave Ellifrit of the Center for Whale Research; Taylor Shedd of The Whale Museum at Friday Harbor; Deborah Giles, science and research director for the nonprofit Wild Orca and researcher at the UW Center for Conservation Biology; and Brad Hanson of NOAA's Northwest Fisheries Science Center.

I thank scientists Marla Holt and Mike Ford, also of NOAA's Northwest Fisheries Science Center, for taking so much time to share their knowledge with me. I thank Tom Quinn, fisheries biologist at the University of Washington, for imparting his expertise and perspective on salmon. I'm grateful to Andrew Trites of the University of British Columbia, who encouraged me to study the loss of salmon in California to better understand the southern resident orcas' plight.

Paul Spong and Helena Symonds opened OrcaLab to me, and Rob Williams and Erin Ashe of Oceans Initiative introduced me to the world of the northern residents and shared their insights built from their work with orcas on both sides of the US-Canada border. John K. B. Ford generously told his stories of his early work that revealed the society of resident orcas and their learned dialects. Joe Gaydos, science director at the SeaDoc Society, took me twice on the *Molly B* for crucial reporting trips, sharing his wonder for the Salish Sea and its creatures. Scientists at the Friday Harbor Labs welcomed me into their special world and provided key interviews. Jonathan Ambrose of NOAA's Central Valley office taught me the history of California winter-run chinook. Doug Killam of the California Department of Fish and Wildlife spent a long, hot day on the water showing me the most extreme environment I've ever seen salmon try to live in.

Stephanie Buffum, former executive director of the nonprofit Friends of the San Juans, helped and encouraged me especially in learning about forage fish. Ted Griffin didn't have to talk to me about capturing orcas but did so and generously shared his personal photos and home movies, hugely enriching the telling of this important chapter in our relationship with orcas. Historian Jason Colby of the University of Victoria was a wealth of knowledge on the capture era. Anne Shaffer of the Coastal Watershed Institute got me in the water to see the magnificence of forage fish in the near shore. Mel Elofson at the Lower Elwha Klallam Tribe was the most knowledgeable guide I could have asked for in understanding the meaning and scope of the Elwha restoration. To elders and leaders at the Lummi Nation, especially the late chief Tsi'li'xw Bill James, I owe a special debt. Historian Coll Thrush helped with crucial insights about the unreconciled history of central Puget Sound.

My thanks to Director Megan Dethier of the University of Washington Friday Harbor Labs for providing the perfect place to write this manuscript.

Many scientists reviewed this manuscript for accuracy, generously taking time from their own work to help me with mine. Any mistakes are of course my own.

To my colleagues at *The Seattle Times* who lived the first draft of this book with me in countless stories over the years, but especially the "Hostile Waters" newspaper series, thank you for all I learned from you and with you. To schoolteacher Cindy Ebisu and the kids in room 11 at Briarcrest Elementary, thank you for inspiring me and supporting me. To Helen Cherullo and Kate Rogers of Mountaineers Books, thank you for believing in this project. John Howell and Claire Powers shared their fire at their cabin and were wonderful listeners to early chapter drafts.

Finally to M pod, you know who you are, and how you helped. Thank you doesn't begin to say it. But thank you.

ADDITIONAL PHOTOGRAPHY ACKNOWLEDGMENTS

Additional photography has been generously provided by the following individuals and organizations:

Dr. Lance Barrett-Lennard, Ocean Wise Research, Ocean Wise Conservation Association
Nancy Bleck
Center for Whale Research
Dr. John Durban, SEA/Southall Environmental Associates
Explore.org
Dr. Holly Fearnbach, SR3/SeaLife Response Rehab and Research
Katy Laveck Foster, The Whale Sanctuary Project
Dr. Deborah Giles, Wild Orca
John Gussman, Doubleclick Productions
Dr. Martin Haulena, Ocean Wise, Vancouver Aquarium
Dr. Terrell C. Newby
NOAA Fisheries Northwest Fisheries Science Center
Taylor Shedd, Soundwatch, The Whale Museum
Dr. Andrew W. Trites, Institute for the Oceans and Fisheries, University of British Columbia
Washington State Archives

A NOTE ABOUT THE PHOTOGRAPHY

All aerial images of orcas were obtained using remotely controlled drones that were flown non-invasively more than 100 feet above the whales and by permit.

IN GRATITUDE

Thank you to the many donors, including individuals, conservation partners, and foundations whose generosity helped bring this book to life. To learn more about how you can contribute to the ongoing efforts to protect wild places and animals in the Salish Sea, visit www.orca-story.org.

ORCA DEFENDERS $5,000 to $10,000

Tom and Sonia Campion—in honor of Kevin Campion, Amy Campion, and Skyrah Campion Scoggins

The Hugh and Jane Ferguson Foundation

Martha Kongsgaard and Peter Goldman—in honor of Billy Frank Jr.

Greg and Mary Moga

SOUND CHAMPIONS $1,000

Marci and Don Heck
Jane and Joe Jeszeck
In honor of Sandra Peters
Gary Rygmyr and Jennifer Warburton
David Shumate
Tenley Tobin

SALMON STREAMKEEPERS $500

In honor of my daughters, Avery and Maddy
In honor of Liesl and Jeremiah Bogaard
Maria Carney
TJ, Tanya, and Eli King
Joan Poor
Jeff and Sheri Tonn

SALISH STEWARDS $250

In honor of Henry Frederick Bainbridge and Katherine "Lily" Bainbridge
In memory of my Grandma Barbara
Susan Carlson
Helen and Arnie Cherullo
David Claar and Patti Polinsky
In honor of Constantin
Dave Davis and Barb Hardman
Vicki Fagerness
John Leland Gerwin
In honor of Ciela Goodwin
Krista Gordon
In honor of our beautiful grandchildren and their future
In memory of Betty and Harriet Holman
Terri Jordan
In honor of Nancy Kartes
Ellen Kritzman
In honor of Marielle and Erika Larson
In memory of Lola MacDonald
Suzanne L. Mager and Marc S. Pease
In honor of Lynda Mapes
John Ohlson
Orca Conservancy
Kelly Priestley
Mary Helen Riley
Michael Riley
Carollee and Tom Roalkvam
David Secord
Mary Lou Siebert
In honor of Ella and Lily Smits
In honor of the children of the Squaxin Island Tribe
William Stelle
In honor of Tom and Annika Sturgeon
Jesse, Claire, and Helena Taylor

Andrew Urban and Denis Tuzinovic
Janet Wainwright
Charlotte Watts
In honor of Pat and Lloyd Whittall
David L. Workman and Clover K. Lockard

POD PROTECTORS $100

In honor of all those working on behalf of the Salish Sea
In honor of Bob Alsup
Christy Anderson
Liz Banse
In honor of the Robert B. Barnes family
Myst Beal
In honor of Roger Beckett
Elaine Miller Bond
In memory of Brad Borland
Joni L. Bosh and Rob Smith
Jim Burke
Julie Burr
Allda Butler
In honor of Scott Campbell
In honor of the Capoeman-Moffett Family
Betsy Carlson
Amber Carrigan
In honor of Jersey, Journey, Jette, and Jemma Carrigan
Elizabeth and David Christian
Tom Clark and Kate Rogers
Amy Lee Cole
Mike Connor and Christine Werme
Eric Eberhard
Nancy M. Faaren
In honor of Aoife Frost
In gratitude to Ellen and Jim Gamrath
In honor of Howie Garrett
In memory of my father, Wayne Goode
Denise Griffing
In honor of Jordan Haggard
In honor of Ruth Halsell
In honor of Karen Hansen
In honor of Susan Hansen and Fred Noland
David Haskell
Hendrickson Temkin grandchildren
In honor of my grandchildren, James and Bridget Holbrooks
In memory of Wayne Hom
Philip Huang
In honor of my grandfather, Homer B. Hubbard
In honor of Collin, Hadley, and Riley Inman
In memory of Michael J. Kelly
Mary McCauley Kraeszig
In memory of Bernhard and Florence Krummel
Jann Ledbetter
In honor of Al and Flora Leisenring
Tami Livingston
Diana E. M. Lloyd
In memory of Tracy Lord
In memory of Mary Maguire
For Aishwarya and Karishma Mandyam, on their graduation
In honor of Lynda Mapes
In memory of my Aunt Mary
Roger Mellem and Gisela Stehr
In memory of Della Messer
In memory of Joseph Miller
Anne and Vincent Murray
In honor of Native Americans
In honor of Sue Nightingale
In honor of John Pearse
Mark Pearson
Heather Peterson
In memory of Dorothy Powell
Jenny Roman
Vicky M. Semones
Dave Shreffler and Ann Soule
Stephanie Shushan
Amanda and Travis Sipher, in honor of Thea Jordan Sipher
Giulia Good Stefani
In honor of Diane Stuart
In honor of Tahlequah
In honor of Howard Taylor
Diane Telschow
In honor of Tokitae—may she return home soon
In honor of Greg, Sam, and Jack Vamos and Dr. R. M. Turian
Sarah Webb
Susan Yates
Kevin Young
In memory of Zoe

NOTES

INTRODUCTION: TAHLEQUAH

"What is beyond grief?": Deborah Giles (science director for Wild Orca, researcher with the Center for Conservation Biology at University of Washington) interview.

"She was stuck in a loop": Deborah Giles interview.

1. THE PEOPLE THAT LIVE UNDER THE SEA

***Orcinus orca*, "from the realm of the dead"**: Erich Hoyt, *Orca: The Whale Called Killer* (Buffalo, NY: Firefly Books, 2019), 27.

Cultural Bonds

So much of what resident orcas do: John K. B. Ford (Canadian orca expert) interview.

"It was a really exciting time": John K. B. Ford interview.

"The question is why": Darren Croft (Center for Research in Animal Behavior, University of Exeter, UK) interview.

"The amount of salmon in a given year varies": Darren Croft interview.

"J2 was a very old orca": Michael Weiss (field researcher for Croft's lab at the Center for Orca Research, San Juan Island) interview.

A Dwindling Population

more than two hundred orcas: "Southern Resident Killer Whale," Marine Mammal Commission, www.mmc.gov/priority-topics/species-of-concern/southern-resident-killer-whale.

Preferred Prey

In some of his research of prey samples: John K. B. Ford, "Selective Foraging by Fish-Eating Killer Whales *Orcinus Orca* in British Columbia," *Marine Ecology Progress Series*, July 3, 2006.

Subsequent research has confirmed that, in winter, as much as half: Brad Hanson (NOAA Fisheries Northwest Fisheries Science Center) interview; "Species Stock Identification of Prey Consumed by Endangered Southern Resident Killer Whales in their Summer Range," *Endangered Species Research,* March 2010.

Seeing with Sound

It can be hard for humans: Marla Holt (research wildlife biologist with NOAA Fisheries Northwest Fisheries Science Center, Seattle) interview.

Scientists are learning from data collected with these sensors: Marla Holt interview; Jennifer Tennessen (biologist at NOAA Fisheries Northwest Fisheries Science Center, Seattle) interview.

But the orcas will also dive to nearly 1,000 feet: Marla Holt and Jennifer Tennessen, "Kinematic Signatures of Prey Capture from Archival Tags Reveal Sex Differences in Killer Whale Foraging Activity," *Journal of Experimental Biology*, 2019.

"We started to realize these animals were selecting for chinook": John K. B. Ford interview.

"They evolved as chinook specialists": John K. B. Ford interview.

On the Hunt

A group of transients will encircle the sea lion: John K. B. Ford and Graeme M. Ellis, "You Are What You Eat: Foraging Specializations and Their Influence on the Social Organization and Behavior of Killer Whales," in *Primates and Cetaceans Field Research and Conservation of Complex Mammalian Societies*, eds. Juichi Yamagiwa and Leszek Karczmarksi (Tokyo: Springer, 2014); https://doi.org/10.1007/978-4-431-54523-1_4.

2. CAPTIVES

Namu

[Ted] Griffin was a prodigious collector: Ted Griffin interview.

Griffin drummed up publicity with the two local papers: *Seattle Post-Intelligencer* and *The Seattle Times* online archives.

Griffin was given a hero's welcome: Jason Colby (historian at University of Victoria), *Orca: How We Came to Know and Love the Ocean's Greatest Predator* (New York: Oxford University Press, 2018), 79–81.

"They are making music": Ted Griffin quoted in Jason Colby, *Orca*.

"Namu the killer whale": Stanton Patty, "Griffin Rides Namu for the First Time," *The Seattle Times*, October 10, 1965.

"He doesn't want me to go back to shore": Ted Griffin interview.

"Namu holds me hostage for his pleasure": Ted Griffin, *Namu: Quest for the Orca* (Seattle: Gryphon West Publishers, 1982), 196.

published a photo of Griffin: "Making Friends with a Killer Whale," *National Geographic* 129, no. 3, March 1966.

Orcas for Sale

"A pod of killer whales": photo caption "A pod of killer whales rolled beneath a hovering helicopter while Ted Griffin prepared to fire a tranquilizer dart. Two fishing boats and smaller craft were being used in the hunt for the whales," *The Seattle Times*, October 31, 1965; staff photos by Bruce McKim.

"if a tranquilizer harpoon hits a whale": *The Seattle Times*, October 31, 1965 (headline is cut off the clipping).

on a hunt of his own: Merrill Spencer; this incident is relayed by Jason Colby in *Orca*, 89.

Death at Penn Cove

Griffin ordered most of the orcas freed: Ted Griffin interview; Jason Colby, *Orca*, 181.

"They were frantic": Terry Newby interview.

Progressive Animal Welfare Society picketed Griffin's aquarium: *The Seattle Times* photo archives.

Regulating an Orca Live-Capture Fishery

"I had no idea at the time": Ted Griffin interview.

The Last Orca Hunt in America

"It was gruesome": Ralph Munro (staff aide to Governor Dan Evans in 1976, later Washington secretary of state for five terms) interview.

In a news conference: Dan Evans (former Washington governor and US senator) interview.

Lolita

"There was a time not very long ago": Jay Julius (former chairman of the Lummi Nation) interview.

"The misconception about whales": Lori Marino (neuroscientist who leads the Whale Sanctuary Project) interview.

documented the harmful effects of captivity on orcas: Lori Marino, "The Harmful Effects of Captivity and Chronic Stress on the Well-Being of Orcas," *Journal of Veterinary Behavior*, 2019.

"They are so intelligent": Lori Marino interview.

"She was very interested": Keith Morrison (*Dateline* correspondent in Southern California), interview.

"Lolita's tank at the Miami Seaquarium": John Hargrove (former senior trainer at SeaWorld), expert report rendered February 2016, unsealed in a 2018 federal appeals court case.

"From my personal observations": Ingrid Visser (New Zealand–based orca researcher), expert report rendered February 2016, unsealed in a 2018 federal appeals court case.

"kill the Indian and save the man": Richard Henry Pratt (late nineteenth-century former army officer, founder of Carlisle Indian Industrial School, Pennsylvania) http://historymatters.gmu.edu/d/4929.

"We understand what happed to her": Bill James (hereditary chief of the Lummi people) interview.

A Fight About More than One Whale

"My secret recipe": Freddie Lane (former Lummi tribal councilman) interview.

3. HUNGER

The Blob

"What if the frequency": Ritchie Graves (chief of NOAA's Northwest region hydropower division) interview.

Without more food: Robert Lacy, Robert Williams et al., "Evaluating Anthropogenic Threats to Endangered Killer Whales to Inform Effective Recovery Plans," *Nature*, October 2017.

King of Kings

"What makes a fish go": Tom Quinn (fisheries biologist, University of Washington) interview aboard the research vessel *Zephyr*.

A Gift of Abundance

Salmon are a keystone animal and **Even in death, salmon mean wealth**: Jeff Cederholm et al., *Pacific Salmon and Wildlife-Ecological Contexts, Relationships, and Implications for Management*, special ed. Technical Report (Olympia: Washington Department of Fish and Wildlife, 2000).

records of old canneries: Ted Gresh (University of Oregon scientist) et al., "An Estimation of Historic and Current Levels of Salmon Production in the Northeast Pacific Ecosystem: Evidence of a Nutrient Deficit in the Freshwater Systems of the Pacific Northwest," *Fisheries*, January 2011.

Hatcheries Industrialize Salmon

Photos of cannery wharves in Seattle: ca. 1850 Seattle Museum of History and Industry, https://digitalcollections.lib.washington.edu/digital/collection/imlsmohai/id/6987/rec/2.

Fewer Fish, Shorter Seasons

even the fish themselves are smaller: Jan Ohlberger, "Demographic Changes in Chinook Salmon across the Northeast Pacific Ocean," *Fish and Fisheries*, January 2018.

This salmon shrinkage: William Ricker, *Causes of the Decrease in Size and Age of Chinook Salmon*, report for Canadian Technical Report of Fisheries and Aquatic Sciences No. 944, (Nanaimo, B.C.: Fisheries and Oceans Canada, 1980).

So hunger stunts even the ocean's top predator: John Durban and Holly Fearnbach, "Decadal Changes in Adult Size of Salmon-Eating Killer Whales in the Eastern North Pacific," *Endangered Species Research*, January 2019.

the resurgent population of marine mammals: Brandon Chasco et al., "Competing Tradeoffs between Increasing Marine Mammal Predation and Fisheries Harvest of Chinook Salmon," *Nature*, November 20, 2017.

Declining Salmon Diversity

With so much diversity lost: Michael Ford (director of conservation biology division at NOAA Fisheries Northwest Fisheries Science Center, Seattle) interview.

There's no rescue underway: Andrew Trites (professor and director of the Marine Mammal Research Unit at the Institute for Oceans and Fisheries, University of British Columbia) interview.

"all the areas the whales take fish out of": Brad Hanson (research wildlife biologist with NOAA Fisheries Northwest Fisheries Science Center, Seattle) interview aboard the *Tyee III*.

Scat Tracking

no live calf was produced: Sam Wasser (director of the Center for Conservation Biology, University of Washington) interview; "Population Growth Is Limited by Nutritional Impacts on Pregnancy Success in Endangered Southern Resident Killer Whales," *Plos One*, June 29, 2017.

A Fast-Changing Environment

"The environment has changed so quickly": Sheila Thornton, (lead orca scientist for British Columbia's Department of Fisheries and Oceans) interview.

"It's not just the abundance of chinook salmon": Michael Ford interview.

4. THE ROAR BELOW

"We are asking them". Marla Holt interview.

Listening in on the Orcas' Soundscape

The ports of Seattle and Tacoma: Northwest Seaport Alliance, www.nwseaportalliance.com.

the maritime sector employed: Washington State Department of Commerce, www.commerce.wa.gov/growing-the-economy/key-sectors/maritime.

Current demand forecasts: Ocean Shipping Consultants, *Container Traffic Forecast Study—Port of Vancouver 2016*, Vancouver, BC, May 2015, www.portvancouver.com/wp-content/uploads/2015/05/2016-Container-Traffic-Forecast-Study-Ocean-Shipping-Consultants.pdf.

That proposal is under environmental review: environmental assessment, Robert Banks Terminal 2 Project, Vancouver Fraser Port Authority, 2018, www.robertsbankterminal2.com/environmental-assessment/ea-process.

Impacts of Noise

southern residents potentially lose: Vancouver Fraser Port Authority, *Enhancing Cetacean Habitat and Observation Program Study Summary*, Vancouver, BC, May 2017, www.portvancouver.com/wp-content/uploads/2017/01/2017-07-ECHO-Program-Estimating-the-effects-of-noise-from-commercial-vessels-and-whale-watch-boats-on-SRKW.pdf.

large ships have the biggest influence: Scott Veirs (Seattle oceanographer) and Val Veirs (his father and retired physicist on San Juan Island) interviews; Dominic Tollit et al., *Estimating the Effects of Noise from Commercial Vessels and Whale Watch Boats on Southern Resident Killer Whales*, report for the ECHO Program of Vancouver Fraser Port Authority, SMRU Consulting North America, July 2017, https://georgiastrait.org/wp-content/uploads/2018/02/2017-07-ECHO-Program-Estimating-the-effects-of-noise-from-commercial-vessels-and-whale-watch-boats-on-SRKW.pdf.

ships are raising noise levels: Scott Veirs et al., "Ship Noise Extends to Frequencies Used for Echolocation by Endangered Killer Whales," *Environmental Science*, 2016.

Speed is the single most important: Juliana Houghton, Marla Holt et al., "The Relationship between Vessel Traffic and Noise Levels Received by Killer Whales," *Plos One*, December 2015.

a surge in humpback whales: Jeff Friedman (US president of Pacific Whale Watch Association) interview.

When Threats Combine

"It was like a village out there": Ken Balcomb (founding director of the nonprofit Center for Whale Research, San Juan Island, www.whaleresearch.com) interview.

it depends not only on the noise: Rob Williams (chief scientist, Oceans Initiative) interview.

the southern resident orcas they monitored: David Lusseau, Rob Williams et al., "Vessel Traffic Disrupts the Foraging Behavior of Southern

Resident Killer Whales," *Endangered Species Research*, 2009.

the areas where orcas can effectively communicate: Rob Williams et al., "Acoustic Quality of Critical Habitats for Three Threatened Whale Populations," *Animal Conservation*, 2013.

5. HOSTILE WATERS

want to raise the dam higher: US Bureau of Reclamation California–Great Basin, "Shasta Dam and Reservoir Enlargement Project," October 2018, www.usbr.gov/mp/ncao/shasta-enlargement.html.

The Central Valley's Thirsty Crops

With winter-run chinook headed to extinction: NOAA Fisheries, "Species in the Spotlight: Priority Actions, 2016–2020, Sacramento River Winter-Run Chinook Salmon, *Oncorhynchus tshawytscha*," NOAA, https://repository.library.noaa.gov/view/noaa/10746.

Two studies of winter chinook: NOAA Fisheries, "Biological Opinion and Conference Opinion on the Long-Term Operations of the Central Valley Project and State Water Project," NOAA, June 4, 2009, www.fisheries.noaa.gov/resource/document/biological-opinion-and-conference-opinion-long-term-operations-central-valley; "Biological Opinion for the Long-Term Operation of the CVP and SWP Endangered Species Act Section 7(a)(2) Biological Opinion Reinitiation of Consultation on the Long-Term Operation of the Central Valley Project and the State Water Project NMFS Consultation Number: WCR-2019-11484," https://ca-times.brightspotcdn.com/9e/28/988004cc4c59aa15c6ab9f18e1e4/nmfs-jeopardy-biop-2019-ocr.pdf.

"We were so close": Jonathan Ambrose (reintroduction coordinator for NOAA's National Marine Fisheries Service, Central Valley office, Sacramento) interview.

all the chinook salmon he could catch: J. Parker Whitney, "Salmon Fishing off Monterey," *Sunset Magazine*, January 1903.

Battling Extinction on the Columbia and Snake

the southern residents swimming: Northwest Fisheries Science Center, "Southern Resident Killer Whale Tagging," NOAA Fisheries, n.d., www.nwfsc.noaa.gov/research/divisions/cb/ecosystem/marinemammal/satellite_tagging/index.cfm.

I once rode one of these fish barges: US Army Corps of Engineers, "Fish Programs," www.nww.usace.army.mil/Missions/Fish-Programs.

an organic machine: Richard White, *The Organic Machine: The Remaking of the Columbia River* (New York: Hill and Wang, 1995), 104.

Seattle's Only River

more than a hundred years' war on this river: Mike Sato, *The Price of Taming a River* (Seattle: Mountaineers Books, 1997), 37–38.

oversaw the creation of two cuts: Hiram M. Chittenden (US Army Corps of Engineers), www.nws.usace.army.mil/Missions/Civil-Works/Locks-and-Dams/Chittenden-Locks/; David B. Williams, Jennifer Ott, and History Link staff, *Waterway: The Story of Seattle's Locks and Ship Canal* (Seattle: History Link, 2017).

"That was quite a day": Joseph Moses, Duwamish descendant, August 28, 1916, www.historylink.org/File/686.

"More land for industry": Howard Hanson (engineer), "More Land for Industry, the Story of Flood Control in the Green River Valley," *Pacific Northwest Quarterly*, 1957.

Remaking the Duwamish

A $342 million seventeen-year Superfund cleanup: Superfund Home, "Lower Duwamish Waterway, Seattle, WA," US Environmental Protection Agency, n.d., https://cumulis.epa.gov/supercpad/cursites/csitinfo.cfm?id=1002020.

93 acres of habitat restoration: Port of Seattle, "Duwamish Cleanup," 2019, www.portseattle.org/projects/duwamish-cleanup.

Not to mention ciprofloxacin: Jim Meador (environmental toxicologist at NOAA's Northwest Fisheries Science Center, Seattle) et al., "Contaminants of Emerging Concern in a Large Temperate Estuary," *Environmental Pollution*, June 2016.

An article I wrote about the study: "Drugs Found in Puget Sound Salmon from Tainted Wastewater," *The Seattle Times*, February 23, 2016.

maybe the only time my reporting: Stephen Colbert, *The Late Show*, CBS, March 29, 2016, www.youtube.com/watch?v=X1WtkwWsQyA.

Green River's Farm Fields Turned to Warehouses

"The only controversy is why it is taking so long": Fred Goetz (Endangered Species Act coordinator for the US Army Corps of Engineers' Seattle district office) interview at Howard Hanson Dam.

Survivors

They didn't matter to the settlers: Coll Thrush (historian at University of British Columbia), interview about his article "City of the Changers," *Pacific Historical Review*, February 2006.

"The people who did all this": Coll Thrush, interview.

A Race against Time

That's both good and bad news: Jeff Davis (director of conservation for the Washington Department of Fish and Wildlife), interview.

"We haven't gone far enough": Ron Warren (director of fish programs for the Washington Department of Fish and Wildlife), interview.

6. A LAND TO THE NORTH

"The northern resident population was persecuted": John K. B. Ford interview.

Taking them from their families was particularly harmful: Rob Williams and David Lusseau, "A Killer Whale Social Network Is Vulnerable to Targeted Removals," *Biology Letters*, January 2007.

The northern resident population has grown: John K. B. Ford interview.

using underwater listening devices: John K. B. Ford interview.

only one dip in the population since 2001: Fisheries and Oceans Canada, "Killer Whale (Northeast Pacific Resident Population)," Government of Canada, December 6, 2018, www.dfo-mpo.gc.ca/species-especes/profiles-profils/killerWhaleNorth-PAC-NE-epaulardnord-eng.html.

Skana

thousands of hours of orca sounds and images: Explore.org, "OrcaLab Main Cams," https://explore.org/livecams/orcas/orcalab-base; Orca-Live.net, "OrcaLab Base," www.orca-live.net/community/index.html.

An Orca Sanctuary

Drone photography has revealed: Andrew Trites interview.

underwater footage: BC Parks, "Robson Bight (Michael Bigg) Ecological Reserve," http://bcparks.ca/eco_reserve/robsonb_er.html.

What Works There Is Needed Here Too

PCB effects on reproduction: Jean-Pierre Desforges et al., "Predicting Global Killer Whale Population Collapse from PCB Pollution," *Science*, September 26, 2018.

7. THE WORK

The Grand Experiment

recording the condition of the river: Jeffrey J. Duda et al. (interdisciplinary team of scientists from the Lower Elwha Klallam Tribe, academia, nonprofits, and federal and state agencies), "Science Partnership between US Geological Survey and the Lower Elwha Klallam Tribe—Understanding the Elwha River Dam Removal Project," USGS Fact Sheet 2018-3025, April 16, 2018, https://doi.org/10.3133/fs20183025.

Dam removal increased by two orders of magnitude: Jonathan Warrick (US Geological Survey) et al., "World's Largest Dam Removal Reverses Coastal Erosion," *Nature*, September 27, 2019.

so astutely observed in his quirky, wonderful blog: Ian Miller (coastal hazard specialist for Washington SeaGrant), *Coast Nerd Gazette*, http://coastnerd.blogspot.com.

"After the dams were removed": Matt Beirne (natural resources director for the Lower Elwha Klallam Tribe) interview.

A Cautionary Tale

assessing the cumulative effects of bulkheads: Megan Dethier et al., "Multiscale Impacts of Armoring on Salish Sea Shorelines: Evidence for Cumulative and Threshold Effects," *Estuarine, Coastal and Shelf Science*, June 2016.

"The concept of saving the orcas": Megan Dethier interview at Friday Harbor Labs.

Kelp also is a fundamental food source: David Duggins (supervisor of marine operations and senior research scientist at University of Washington's Friday Harbor Labs, San Juan Island), interview; D. O. Duggins et al., "Magnification of Secondary Production by Kelp Detritus in Coastal Marine Ecosystems," *Science*, 1989.

Elwha Resurgent

A host of animals are colonizing habitat: Rebecca McCaffery (USGS Forest and Rangeland Ecosystem Science Center's Olympic Field Station, Port Angeles) interview; Rebecca McCaffery et al., "Terrestrial Fauna Are Agents and Endpoints in Ecosystem Restoration Following Dam Removal," *Ecological Restoration*, June 2018.

That is boosting juvenile salmon growth: Chris Tonra (professor of avian wildlife ecology at Ohio State University) et al., "The Rapid Return of Marine Derived Nutrients to a Freshwater Food Web Following Dam Removal," *Biological Conservation*, August 2015.

Return of the Kings

"It's the most exciting discovery I have had in science": John McMillan (science director for Trout Unlimited's wild steelhead initiative) interview.

sampled fish diets before and after dam removal: Sarah Morley (NOAA Fisheries Northwest Fisheries Science Center, Seattle) interview; Sarah Morley et al., "Shifting Food Web Structure during Dam Removal—Disturbance and Recovery during a Major Restoration Action," *Plos One*, 15(9): e0239198, https://doi.org/10.1371/journal.pone.0239198.

BIBLIOGRAPHY

The Capture Era

Colby, Jason. *Orca: How We Came to Know and Love the Ocean's Greatest Predator*. New York: Oxford University Press, 2018.

Griffin, Ted. *Namu: Quest for the Killer Whale*. Seattle: Gryphon West Publishers, 1982.

McKim, Bruce, photographer. *The Seattle Times,* October 31, 1965.

National Geographic Society. "Making Friends with a Killer Whale." *National Geographic* 129, no. 3 (March 1966).

Patty, Stanton. "Griffin Rides Namu for the First Time." *Seattle Times,* October 10, 1965.

Pollard, Sandra. *A Puget Sound Orca in Captivity*. Charleston, SC: The History Press, 2019.

———. *Puget Sound Whales for Sale*. Charleston, SC: The History Press, 2014.

Natural History and Behavior of Orcas

BC Parks. "Robson Bight (Michael Bigg) Ecological Reserve." British Columbia Provincial Government. http://bcparks.ca/eco_reserve/robsonb_er.html,

Cederholm, Jeff, et al. *Pacific Salmon and Wildlife Ecological Contexts, Relationships, and Implications for Management*. Special Edition Technical Report. Olympia: Washington Department of Fish and Wildlife, 2000.

Chasco, Brandon, et al. "Competing Tradeoffs between Increasing Marine Mammal Predation and Fisheries Harvest of Chinook Salmon." *Nature*, November 20, 2017.

Desforges, Jean-Pierre, et al. "Predicting Global Killer Whale Population Collapse from PCB Pollution," *Science* 361, no. 6409 (September 26, 2018). https://doi.org/10.1126/science.aat1953.

Durban, John, and Holly Fearnbach. "Decadal Changes in Adult Size of Salmon-Eating Killer Whales in the Eastern North Pacific." *Endangered Species Research*, January 2019.

Explore.org. "OrcaLab Main Cams." Explore.org. https://explore.org/livecams/orcas/orcalab-base.

Fisheries and Oceans Canada. "Killer Whale (Northeast Pacific Resident Population)." Government of Canada, December 6, 2018. www.dfo-mpo.gc.ca/species-especes/profiles-profils/killerWhaleNorth-PAC-NE-epaulardnord-eng.html.

Ford, John K. B. *Marine Mammals of British Columbia*. Victoria: Royal British Columbia Museum, 2014.

———. *Transients: Mammal-Hunting Killer Whales*. Vancouver: University of British Columbia Press, 1999.

Ford, John K. B., and Graeme M. Ellis. "You Are What You Eat: Foraging Specializations and Their Influence on the Social Organization and Behavior of Killer Whales." In *Primates and Cetaceans: Field Research and Conservation of Complex Mammalian Societies*, eds. Juichi Yamagiwa and Leszek Karczmarski, ch. 4. Tokyo: Springer, 2014. https://doi.org/10.1007/978-4-431-54523-1_4.

Ford, John K. B., Graeme M. Ellis, and Kenneth C. Balcomb. *The Natural History and Genealogy of* Orcinus Orca *in British Columbia and Washington State*. Seattle: University of Washington Press, 2000.

Gresh, Ted, et al. "An Estimation of Historic and Current Levels of Salmon Production in the Northeast Pacific Ecosystem: Evidence of a Nutrient Deficit in the Freshwater Systems of the Pacific Northwest." *Fisheries,* January 2011.

Haig-Brown, Roderick. *A River Never Sleeps*. New York: Skyhorse, 2010.

Hanson, Brad. "Species Stock Identification of Prey Consumed by Endangered Southern Resident Killer Whales in their Summer Range." *Endangered Species Research*, March 2010.

Holt, Marla, and Jennifer Tennessen. "Kinematic Signatures of Prey Capture from Archival Tags Reveal Sex Differences in Killer Whale Foraging Activity." *Experimental Biology*, 2019.

Hoyt, Erich. *Orca: The Whale Called Killer*. Buffalo, NY: Firefly Books, 2019.

Lacy, Robert, Robert Williams et al. "Evaluating Anthropogenic Threats to Endangered Killer Whales to Inform Effective Recovery Plans." *Nature*, October 2017.

Mapes, Lynda. "Hostile Waters." *The Seattle Times,* November 2018–September 2019.

Marino, Lori. "The Harmful Effects of Captivity and Chronic Stress on the Well-Being of Orcas." *Journal of Veterinary Behavior,* 2019.

Neiwert, David. *Of Orcas and Men: What Killer Whales Can Teach Us*. New York: Overlook Press, 2016.

Ohlberger, Jan. "Demographic Changes in Chinook Salmon across the Northeast Pacific Ocean," *Fish and Fisheries*, January 2018.

Orca-Live.net. "OrcaLab Base." Orca-Live.net. www.orca-live.net/community/index.html.

Safina, Carl. *Beyond Words: What Animals Think and Feel*. New York: Henry Holt and Co., 2015.

Sato, Mike. *The Price of Taming a River*. Seattle: Mountaineers Books, 1997.

Wasser, Sam. "Population Growth Is Limited by Nutritional Impacts on Pregnancy Success in Endangered Southern Resident Killer Whales." *Plos One*, June 29, 2017.

Williams, Rob, and David Lusseau. "A Killer Whale Social Network Is Vulnerable to Targeted Removals." *Biology Letters*, January 2007.

Salish Sea Ecology

DeLella Benedict, Audrey, and Joseph Carl Gaydos. *The Salish Sea: Jewel of the Pacific Northwest*. Seattle: Sasquatch Books, 2015.

Houghton, Juliana, Marla Holt et al. "The Relationship between Vessel Traffic and Noise Levels Received by Killer Whales." *Plos One*, December 2015.

Lusseau, David, Rob Williams et al. "Vessel Traffic Disrupts the Foraging Behavior of Southern Resident Killer Whales." *Endangered Species Research*, 2009.

Ocean Shipping Consultants. *Container Traffic Forecast Study—Port of Vancouver 2016*. Vancouver Fraser Port Authority. Vancouver, BC, May 2015. www.portvancouver.com/wp-content/uploads/2015/05/2016-Container-Traffic-Forecast-Study-Ocean-Shipping-Consultants.pdf.

Tollit, Dominic, et al. *Estimating the Effects of Noise from Commercial Vessels and Whale Watch Boats on Southern Resident Killer Whales*. Report for the ECHO Program of Vancouver Fraser Port Authority, SMRU Consulting North America, July 2017. https://georgiastrait.org/wp-content/uploads/2018/02/2017-07-ECHO-Program-Estimating-the-effects-of-noise-from-commercial-vessels-and-whale-watch-boats-on-SRKW.pdf.

Vancouver Fraser Port Authority. *Enhancing Cetacean Habitat and Observation Program Study Summary*. Vancouver Fraser Port Authority, Vancouver, BC, May 2017. www.portvancouver.com/wp-content/uploads/2017/01/2017-07-ECHO-Program-Estimating-the-effects-of-noise-from-commercial-vessels-and-whale-watch-boats-on-SRKW.pdf.

———. *Environmental Assessment, Robert Banks Terminal 2 Project*. Vancouver Fraser Port Authority. Vancouver, BC, 2018. www.robertsbankterminal2.com/environmental-assessment/ea-process/.

Veirs, Scott, et al. "Ship Noise Extends to Frequencies Used for Echolocation by Endangered Killer Whales." *Environmental Science*, 2016.

Williams, Rob, et al. "Acoustic Quality of Critical Habitats for Three Threatened Whale Populations." *Animal Conservation*, 2013.

Salmon

Dethier, Megan, et al. "Multiscale Impacts of Armoring on Salish Sea Shorelines: Evidence for Cumulative and Threshold Effects." *Estuarine, Coastal and Shelf Science*, June 2016.

Duda, Jeffrey J., et al. "Science Partnership between US Geological Survey and the Lower Elwha Klallam Tribe—Understanding the Elwha River Dam Removal Project." USGS Fact Sheet 2018-3025, April 16, 2018. https://doi.org/10.3133/fs20183025.

Duggins, David O., et al. "Magnification of Secondary Production by Kelp Detritus in Coastal Marine Ecosystems." *Science*, 1989.

Hanson, Howard. "More Land for Industry, the Story of Flood Control in the Green River Valley." *Pacific Northwest Quarterly*, 1957.

Mapes, Lynda. "Drugs Found in Puget Sound Salmon from Tainted Wastewater." *The Seattle Times*, February 23, 2016.

McCaffery, Rebecca, et al. "Terrestrial Fauna Are Agents and Endpoints in Ecosystem Restoration Following Dam Removal." *Ecological Restoration*, June 2018.

Meador, Jim, et al. "Contaminants of Emerging Concern in a Large Temperate Estuary." *Environmental Pollution*, June 2016.

Morley, Sarah, et al. "Shifting Food Web Structure during Dam Removal—Disturbance and Recovery during a Major Restoration Action." *Plos One*, *E* 15(9):e0239198. https://doi.org/10.1371/journal.pone.0239198.

NOAA Fisheries. "Biological Opinion for the Long-Term Operation of the CVP and SWP Endangered Species Act Section 7(a)(2) Biological Opinion Reinitiation of Consultation on the Long-Term Operation of the Central Valley Project and the State Water Project NMFS Consultation Number: WCR-2019-11484." NOAA. https://ca-times.brightspotcdn.com/9e/28/988004cc4c59aa15c6ab9f18e1e4/nmfs-jeopardy-biop-2019-ocr.pdf.

———. "Biological Opinion and Conference Opinion on the Long-Term Operations of the Central Valley Project and State Water Project." NOAA, June 4, 2009. www.fisheries.noaa.gov/resource/document/biological-opinion-and-conference-opinion-long-term-operations-central-valley.

———. "Species in the Spotlight: Priority Actions, 2016–2020, Sacramento River Winter-Run Chinook Salmon, *Oncorhynchus tshawytscha*." NOAA. https://repository.library.noaa.gov/view/noaa/10746.

Northwest Fisheries Science Center. "Southern Resident Killer Whale Tagging." NOAA Fisheries, n.d. www.nwfsc.noaa.gov/research/divisions/cb/ecosystem/marinemammal/satellite_tagging/index.cfm.

Port of Seattle. "Duwamish Cleanup." Port of Seattle, 2019. www.portseattle.org/projects/duwamish-cleanup.

Ricker, William. "Causes of the Decrease in Size and Age of Chinook Salmon." Canadian Technical Report of Fisheries and Aquatic Sciences No. 944. Nanaimo, B.C.: Fisheries and Oceans Canada, 1980.

Superfund Home. "Lower Duwamish Waterway, Seattle, WA." US Environmental Protection Agency, n.d. https://cumulis.epa.gov/supercpad/cursites/csitinfo.cfm?id=1002020.

Thrush, Coll. "City of the Changers." *Pacific Historical Review*, February 2006.

Tonra, Chris, et al. "The Rapid Return of Marine-Derived Nutrients to a Freshwater Food Web Following Dam Removal." *Biological Conservation*, August 2015.

US Bureau of Reclamation California–Great Basin. "Shasta Dam and Reservoir Enlargement Project." US Bureau of Reclamation, October 2018. www.usbr.gov/mp/ncao/shasta-enlargement.html.

Warrick, Jonathan, et al. "World's Largest Dam Removal Reverses Coastal Erosion." *Nature*, September 27, 2019.

White, Richard. *The Organic Machine: The Remaking of the Columbia River*. New York: Hill and Wang, 1995.

Whitney, J. Parker. "Salmon Fishing off Monterey." *Sunset Magazine*, January 1903.

Williams, David B., Jennifer Ott, and History Link staff. *Waterway: The Story of Seattle's Locks and Ship Canal*. Seattle: History Link, 2017.

RESOURCES

Center for Whale Research, San Juan Island, Washington: Natural history, photo catalog of the southern resident orcas with sightings throughout the year, and useful research and demographic information. www.whaleresearch.com/

National Oceanic and Atmospheric Administration: Natural history, status, and recovery plans for the southern resident orcas. www.fisheries.noaa.gov/west-coast/endangered-species-conservation/southern-resident-killer-whale-orcinus-orca

The SeaDoc Society: Excellent mini-documentaries about the Salish Sea habitat and educational resources. www.seadocsociety.org/

The Seattle Times: Special report on the southern resident orca extinction crisis, including maps, graphics, and mini-documentaries. www.seattletimes.com/seattle-news/environment/hostile-waters-orcas-killer-whale-puget-sound-washington-canada/

Washington State Archives: Outstanding resource for historic photos of capture era and documentation of capture policy. www.sos.wa.gov/archives/

The Whale Museum, Friday Harbor, Washington: Natural history and photographic catalog of the southern resident orcas. https://whalemuseum.org/

The Whale Trail: Guidance on land-based whale watching to avoid disturbance of the southern resident orcas in their habitat; maps and resources available. https://thewhaletrail.org/

ABOUT THE AUTHOR

Lynda V. Mapes is a reporter at *The Seattle Times*, where she specializes in coverage of the environment. She is a two-time team award winner of the international Kavli Gold Award for science journalism from the American Association for the Advancement of Science, for her reporting on orcas (2019) and dam removal on the Elwha River (2012). She also was part of the team that won top honors in 2019 from the national Online News Association as well as the National Headliner Award for *The Seattle Times* series, "Hostile Waters: Orcas in Peril." The Western Washington Chapter of the Society of Professional Journalists named Lynda journalist of the year in 2019, and NOAA Fisheries recognized her in 2016 with the prestigious Dr. Nancy Foster Habitat Conservation Award.

In 2013–2014 Lynda was awarded a nine-month Knight fellowship in Science Journalism at Massachusetts Institute of Technology. In 2014–2015, she was a Bullard Fellow at the Harvard Forest, exploring the human and natural history of a single one-hundred-year-old oak.

In addition to her staff position at *The Seattle Times*, Lynda has authored five books. She lives in Seattle.

PRIMARY PHOTOGRAPHER

Steve Ringman has been a photographer for *The Seattle Times* for over twenty-five years. His passion is for stories involving environmental issues including climate change, fisheries, and forestry that have taken him around the world. He has been twice named national Newspaper Photographer of the Year by the National Press Photographers; his other awards include the Knight-Risser Prize for Western Environmental Journalism, and he is part of the multiple award-winning team behind "Hostile Waters: Orcas in Peril," published by *The Seattle Times*.

ABOUT *THE SEATTLE TIMES*

The Seattle Times is proud to be one of the few remaining independent and locally owned metropolitan news media organizations in the United States and the most-visited digital information source in Washington State. As the region's most trusted news media company, dedicated to public service, *The Seattle Times* serves the Northwest with thoughtful, Pulitzer Prize–winning journalism and tells the uniquely local stories you won't find anywhere else.

Founded in 1896 by Alden J. Blethen, *The Seattle Times* remains a family-owned business, now led by the Blethen family's fourth and fifth generations. *The Seattle Times* is deeply rooted in the community with a steadfast commitment to its mission: to provide principled, investigative news coverage and the highest standards of public-service journalism to the Northwest and beyond.

The Seattle Times has also emerged as a pioneer among news organizations for groundbreaking new funding models to secure the future of the free press. Its community-funded, public-service journalism projects go beyond reporting to explore viable solutions to urgent local issues. *Seattle Times* journalism has impacted public policy and resulted in action in local, county and state government.

The Seattle Times is the winner of eleven Pulitzer Prizes, including the 2020 Pulitzer Prize for National Reporting for yearlong coverage that exposed design flaws in Boeing's 737 Max. *The Seattle Times* also won the international American Association for the Advancement of Science (AAAS) Kavli Science Journalism Award for a special report about the plight of endangered Pacific Northwest resident orcas. Environmental journalism is a priority as *The Seattle Times* devotes resources to journalism initiatives focused on issues vital to the future of our region and community, including ocean acidification, destruction of global forest ecosystems, and our warming planet, a defining issue of our time.

BRAIDED RIVER

BRAIDED RIVER, the conservation imprint of Mountaineers Books, combines photography and writing to bring a fresh perspective to key environmental issues facing western North America's wildest places. Our books reach beyond the printed page as we take these distinctive voices and vision to a wider audience through lectures, exhibits, and multimedia events. Our goal is to build public support for wilderness preservation campaigns, and inspire public action. This work is made possible through the book sales and contributions made to Braided River, a 501(c)(3) nonprofit organization. Please visit BraidedRiver.org. For more information on Orca events and how to contribute to this work, visit www.orca-story.org.

Braided River books may be purchased for corporate, educational, or other promotional sales. For special discounts and information, contact our sales department at 800.553.4453 or mbooks@mountaineersbooks.org.

THE MOUNTAINEERS, founded in 1906, is a nonprofit outdoor activity and conservation organization, whose mission is "to explore, study, preserve, and enjoy the natural beauty of the outdoors . . ." Mountaineers Books supports this mission by publishing travel and natural history guides, instructional texts, and works on conservation and history.

See our website to explore our catalog of 700 outdoor titles:
Mountaineers Books
1001 SW Klickitat Way, Suite 201 • Seattle, WA 98134
800.553.4453 • www.mountaineersbooks.org

Manufactured in China on FSC®-certified paper, using soy-based ink.

First edition, 2021

Braided River Executive Director: Helen Cherullo
Editor in Chief: Kate Rogers
Project Editor: Janet Kimball
Braided River Impact Campaign: Erika Lundahl
Developmental Editor: Linda Gunnarson
Copyeditor: Kris Fulsaas
Cover and Book Designer: Jen Grable
Illustrator: Emily M. Eng, *The Seattle Times*
Cartographer: Erin Greb Cartography
Front cover photo: Southern resident orca, J27 (*Dave Ellifrit/Center for Whale Research; taken under NMFS Permit 21238/DFO SARA 388*)
Back cover photo: K20, a southern resident orca, with Mount Baker in the background (*Steve Ringman,* The Seattle Times)
Endsheets: Mark Nowlin/*The Seattle Times*. Sources: Center for Whale Research; Deborah Giles, Wild Orca and University of Washington Center for Conservation Biology

Frontispiece: The thrill of orcas leaping from clear, cold seas is a signature of the Pacific Northwest region. *(Mark Malleson/Center for Whale Research; taken under NMFS Permit 21238 and DFO SARA Permit 388)*; **Page 4:** Chief Tsi'li'xw Bill James of the Lummi Nation *(Ken Lambert/*The Seattle Times*)* **Page 8:** Orcas frequently spy-hop to look around. They have excellent vision both underwater and above and can even look behind themselves as they swim. *(Dr. Deborah Giles/Center for Whale Research; taken under NMFS Permit 15569 and DFO SARA Permit 288)*; **Page 180:** Three new orca babies have been born to the southern residents since January 2019, including L124, pictured with its mother, L77. Mother orca Tahlequah also gave birth to a healthy male calf on September 4, 2020. *(Mark Malleson/Center for Whale Research; taken under NMFS Permit 21238 and DFO SARA Permit 388)*; **Page 192:** Orca societies have coexisted across the world's oceans for millions of years. Whether southern resident orcas and humans can continue to coexist, too, is yet to be seen. *(Ken Balcomb/Center for Whale Research; taken under NMFS Permit 15569)*

Library of Congress Cataloging-in-Publication Data is on file for this title at at https://lccn.loc.gov/2020039223

ISBN 978-1-68051-326-4